A SHORTER COURSE OF THEORETICAL PHYSICS

VOLUME 1

MECHANICS AND ELECTRODYNAMICS

A SHORTER COURSE OF THEORETICAL PHYSICS

VOLUME 1

MECHANICS AND ELECTRODYNAMICS

BY

L. D. LANDAU

AND

E. M. LIFSHITZ

Institute of Physical Problems, U.S.S.R. Academy of Sciences

TRANSLATED FROM THE RUSSIAN BY
J. B. SYKES AND M. HAMERMESH

PERGAMON PRESS

OXFORD · NEW YORK · TORONTO
SYDNEY · BRAUNSCHWEIG

Pergamon Press Offices:

U.K.	Pergamon Press Ltd., Headington Hill Hall, Oxford, England
U.S.A.	Pergamon Press Inc., Maxwell House, Fairview Park, Elmsford, New York 10523, U.S.A.
CANADA	Pergamon of Canada Ltd., 207 Queen's Quay West, Toronto 1, Canada
AUSTRALIA	Pergamon Press (Aust.) Pty. Ltd., 19a Boundary Street, Rushcutters Bay, N.S.W. 2011, Australia
FRANCE	Pergamon Press SARL, 24 rue des Ecoles, 75240 Paris, Cedex 05, France
WEST GERMANY	Pergamon Press GmbH, 3300 Braunschweig, Postfach 2923, Burgplatz 1, West Germany

First edition 1972

Reprinted 1975

Library of Congress Catalog Card No. 74-167927

Printed in Great Britain by
Biddles Ltd., Guildford, Surrey

ISBN 0 08 016739 X

CONTENTS

5. SMALL OSCILLATIONS

6. MOTION OF A RIGID BODY

7. THE CANONICAL EQUATIONS

8. THE PRINCIPLE OF RELATIVITY

9. RELATIVISTIC MECHANICS

Part II. Electrodynamics

10. CHARGES IN ELECTROMAGNETIC FIELDS

PREFACE

Despite our attempts to be rigorously selective, the volumes of the *Course of Theoretical Physics* grow larger at every revision. Of course, this is a natural and inevitable consequence of the rapid development of science, but nevertheless the books become thereby less and less convenient for use by students, and indeed by anyone other than those whose profession is theoretical physics.

Faced with this situation, Landau was eagerly considering, in the years preceding his terrible accident, the idea of producing a shorter course of theoretical physics based on the complete course. He felt that the shorter course should comprise the minimum of material that should be familiar to every present-day physicist, working in no matter what branch of physics. The tragedy of 1962 prevented Landau, my teacher and friend, from himself taking part in putting this plan into effect, and the first volume of the course did not appear until after his death.

This shorter course will consist of three volumes:

1. Mechanics and Electrodynamics
2. Quantum Mechanics
3. Macroscopic Physics

The present Volume 1 is essentially a careful abridgement of our *Mechanics* and *The Classical Theory of Fields*. I have tried to take account of all the views expressed by Landau in our early discussions of the plans for these books. In particular, he considered that such a shorter course should not include any discussion of the general theory of relativity. In his opinion, the fundamental physical ideas and results of that theory ought to be described in courses of general physics, while the study of its complete mathematical treatment is necessary (at least for the present) only for theoretical physicists.

The remaining content of Volumes 1 and 2 of the *Course of Theoretical Physics* has been reduced to about one-half. Since the shorter course is not intended to provide a professional knowledge of every technique in theoretical physics, only a small number of problems have been retained as simple illustrations.

May I take this opportunity to express my sincere thanks to Dr. Sykes and Professor Hamermesh for all that they have done in making our books known to English-speaking students of physics. The English editions of the books began with Professor Hamermesh's translation of *The Classical Theory of Fields*, and were continued by those of Dr. Sykes and his colleagues. His prolonged labours have been largely responsible for the carefulness and accuracy of the translations, and his comments have on many occasions helped us to make corrections for the English versions.

E. M. Lifshitz

TRANSLATOR'S NOTE

THE work of Professor Hamermesh and myself as translators of this book has consisted largely in selecting the necessary passages from our previous translations of Volumes 1 and 2 of the same authors' *Course of Theoretical Physics*. Each of us having done this, I then sought to bring the two constituent parts into harmony by eliminating the major differences in notation, orthography and style.

For my own part I wish to acknowledge the valuable advice of Dr. J S. Bell and Dr. J. L. Beeby.

J.B.S.

NOTATION

Mechanical quantities

Generalised coordinates and momenta q_i, p_i

Lagrangian and Hamiltonian functions

(Part I) L, H

(Part II) L, $\mathscr{H}$

Particle energy and momentum

(Part I) E, $\mathbf{p}$

(Part II) $\mathscr{E}$, $\mathbf{p}$

Angular momentum $\mathbf{M}$

Moment of force (torque) $\mathbf{K}$

Inertia tensor I_{ik}

Electromagnetic quantities

Electromagnetic field scalar and vector potentials ϕ, $\mathbf{A}$

Electric and magnetic field intensities $\mathbf{E}$, $\mathbf{H}$

Charge and current densities ϱ, $\mathbf{j}$

Electric and magnetic dipole moments $\mathbf{d}$, $\mathbf{m}$

Mathematical notation

Volume, surface and line elements $\mathrm{d}V$, $\mathrm{d}\mathbf{f}$, $\mathrm{d}\mathbf{l}$

Three-dimensional vector and tensor indices are denoted by Latin letters i, k, l, ..., which take the values x, y, z.

Four-dimensional vector and tensor indices are denoted by Greek letters λ, μ, ν, ..., which take the values 0, 1, 2, 3.

The rule for raising and lowering four-dimensional indices is given on page 130.

The rule of summation over repeated (dummy) indices is given on pages 81 and 130.

PART I

MECHANICS

CHAPTER 1

THE EQUATIONS OF MOTION

§1. Generalised coordinates

One of the fundamental concepts of mechanics is that of a *particle*. By this we mean a body whose dimensions may be neglected in describing its motion. The possibility of so doing depends, of course, on the conditions of the problem concerned. For example, the planets may be regarded as particles in considering their motion about the Sun, but not in considering their rotation about their axes.

The position of a particle in space is defined by its position vector $\mathbf{r}$, whose components are its Cartesian coordinates x, y, z. The derivative $\mathbf{v} = \mathrm{d}\mathbf{r}/\mathrm{d}t$ of $\mathbf{r}$ with respect to the time t is called the *velocity* of the particle, and the second derivative $\mathrm{d}^2\mathbf{r}/\mathrm{d}t^2$ is its *acceleration*. In what follows we shall, as is customary, denote differentiation with respect to time by placing a dot above a letter: $\mathbf{v} = \dot{\mathbf{r}}$.

To define the position of a system of N particles in space, it is necessary to specify N position vectors, i.e. $3N$ coordinates. The number of independent quantities which must be specified in order to define uniquely the position of any system is called the number of *degrees of freedom*; here, this number is $3N$. These quantities need not be the Cartesian coordinates of the particles, and the conditions of the problem may render some other choice of coordinates more convenient. Any s quantities $q_1, q_2, \ldots, q_s$ which completely define the position of a system with s degrees of freedom are called *generalised coordinates* of the system, and the derivatives $\dot{q}_i$ are called its *generalised velocities*.

When the values of the generalised coordinates are specified, however, the "mechanical state" of the system at the instant considered is not yet determined in such a way that the position of the system at subsequent instants can be predicted. For given values of the coordinates, the system can have any velocities, and these affect the position of the system after an infinitesimal time interval dt.

If all the coordinates and velocities are simultaneously specified, it is known from experience that the state of the system is completely determined and that its subsequent motion can, in principle, be calculated. Mathematically, this means that, if all the coordinates q and velocities $\dot{q}$ are given at some instant, the accelerations $\ddot{q}$ at that instant are uniquely defined.†

The relations between the accelerations, velocities and coordinates are called the *equations of motion.* They are second-order differential equations for the functions $q(t)$, and their integration makes possible, in principle, the determination of these functions and so of the path of the system.

§2. The principle of least action

The most general formulation of the law governing the motion of mechanical systems is the *principle of least action* or *Hamilton's principle*, according to which every mechanical system is characterised by a definite function $L(q_1, q_2, \ldots, q_s, \dot{q}_1, \dot{q}_2, \ldots, \dot{q}_s, t)$, or briefly $L(q, \dot{q}, t)$, and the motion of the system is such that a certain condition is satisfied.

Let the system occupy, at the instants t_1 and t_2, positions defined by two sets of values of the coordinates, $q^{(1)}$ and $q^{(2)}$. Then the condition is that the system moves between these positions in such a way that the integral

$$S = \int_{t_1}^{t_2} L(q, \dot{q}, t)\, \mathrm{d}t \tag{2.1}$$

† For brevity, we shall often conventionally denote by q the set of all the coordinates $q_1, q_2, \ldots, q_s$, and similarly by $\dot{q}$ the set of all the velocities.

takes the least possible value. The function L is called the *Lagrangian* of the system concerned, and the integral (2.1) is called the *action.*

The fact that the Lagrangian contains only q and $\dot{q}$, but not the higher derivatives $\ddot{q}$, $\dddot{q}$, etc., expresses the result already mentioned, that the mechanical state of the system is completely defined when the coordinates and velocities are given.

Let us now derive the differential equations which solve the problem of minimising the integral (2.1). For simplicity, we shall at first assume that the system has only one degree of freedom, so that only one function $q(t)$ has to be determined.

Let $q = q(t)$ be the function for which S is a minimum. This means that S is increased when $q(t)$ is replaced by any function of the form

$$q(t)+\delta q(t), \tag{2.2}$$

where $\delta q(t)$ is a function which is small everywhere in the interval of time from t_1 to t_2; $\delta q(t)$ is called a *variation* of the function $q(t)$. Since, for $t = t_1$ and for $t = t_2$, all the functions (2.2) must take the values $q^{(1)}$ and $q^{(2)}$ respectively, it follows that

$$\delta q(t_1) = \delta q(t_2) = 0. \tag{2.3}$$

The change in S when q is replaced by $q+\delta q$ is

$$\int_{t_1}^{t_2} L(q+\delta q, \dot{q}+\delta\dot{q}, t)\,dt - \int_{t_1}^{t_2} L(q, \dot{q}, t)\,dt.$$

When this difference is expanded in powers of δq and $\delta\dot{q}$ in the integrand, the leading terms are of the first order. The necessary condition for S to have a minimum is that these terms (called the *first variation*, or simply the *variation*, of the integral) should be zero. Thus the principle of least action may be written in the form

$$\delta S = \delta \int_{t_1}^{t_2} L(q, \dot{q}, t)\,dt = 0, \tag{2.4}$$

or, effecting the variation,

$$\int_{t_1}^{t_2} \left(\frac{\partial L}{\partial q}\,\delta q + \frac{\partial L}{\partial \dot{q}}\,\delta\dot{q}\right) dt = 0.$$

Since $\delta\dot{q} = \mathrm{d}\delta q/\mathrm{d}t$, we obtain, on integrating the second term by parts,

$$\delta S = \left[\frac{\partial L}{\partial \dot{q}}\delta q\right]_{t_1}^{t_2} + \int_{t_1}^{t_2}\left(\frac{\partial L}{\partial q} - \frac{\mathrm{d}}{\mathrm{d}t}\frac{\partial L}{\partial \dot{q}}\right)\delta q\,\mathrm{d}t = 0. \tag{2.5}$$

The conditions (2.3) show that the integrated term in (2.5) is zero. There remains an integral which must vanish for all values of δq. This can be so only if the integrand is zero identically. Thus we have

$$\frac{\mathrm{d}}{\mathrm{d}t}\left(\frac{\partial L}{\partial \dot{q}}\right) - \frac{\partial L}{\partial q} = 0.$$

When the system has more than one degree of freedom, the s different functions $q_i(t)$ must be varied independently in the principle of least action. We then evidently obtain s equations of the form

$$\frac{\mathrm{d}}{\mathrm{d}t}\left(\frac{\partial L}{\partial \dot{q}_i}\right) - \frac{\partial L}{\partial q_i} = 0 \qquad (i = 1, 2, \ldots, s). \tag{2.6}$$

These are the required differential equations, called in mechanics *Lagrange's equations.*† If the Lagrangian of a given mechanical system is known, the equations (2.6) give the relations between accelerations, velocities and coordinates, i.e. they are the equations of motion of the system.

Mathematically, the equations (2.6) constitute a set of s second-order equations for s unknown functions $q_i(t)$. The general solution contains $2s$ arbitrary constants. To determine these constants and thereby to define uniquely the motion of the system, it is necessary to know the initial conditions which specify the state of the system at some given instant, for example the initial values of all the coordinates and velocities.

Let a mechanical system consist of two parts A and B which would, if closed, have Lagrangians L_A and L_B respectively. Then, in the limit

† In the calculus of variations they are Euler's equations for the formal problem of determining the extrema of an integral of the form (2.1).

where the distance between the parts becomes so large that the interaction between them may be neglected, the Lagrangian of the whole system tends to the value

$$\lim L = L_A + L_B. \tag{2.7}$$

This additivity of the Lagrangian expresses the fact that the equations of motion of either of the two non-interacting parts cannot involve quantities pertaining to the other part.

It is evident that the multiplication of the Lagrangian of a mechanical system by an arbitrary constant has no effect on the equations of motion. From this, it might seem, the following important property of arbitrariness can be deduced: the Lagrangians of different isolated mechanical systems may be multiplied by different arbitrary constants. The additive property, however, removes this indefiniteness, since it admits only the simultaneous multiplication of the Lagrangians of all the systems by the same constant. This corresponds to the natural arbitrariness in the choice of the unit of measurement of the Lagrangian, a matter to which we shall return in §4.

One further general remark should be made. Let us consider two functions $L'(q, \dot{q}, t)$ and $L(q, \dot{q}, t)$, differing by the total derivative with respect to time of some function $f(q, t)$ of coordinates and time:

$$L'(q, \dot{q}, t) = L(q, \dot{q}, t) + \frac{\mathrm{d}}{\mathrm{d}t} f(q, t). \tag{2.8}$$

The integrals (2.1) calculated from these two functions are such that

$$\begin{aligned} S' &= \int_{t_1}^{t_2} L'(q, \dot{q}, t)\, \mathrm{d}t = \int_{t_1}^{t_2} L(q, \dot{q}, t)\, \mathrm{d}t + \int_{t_1}^{t_2} \frac{\mathrm{d}f}{\mathrm{d}t}\, \mathrm{d}t \\ &= S + f(q^{(2)}, t_2) - f(q^{(1)}, t_1), \end{aligned}$$

i.e. they differ by a quantity which gives zero on variation, so that the conditions $\delta S' = 0$ and $\delta S = 0$ are equivalent, and the form of the equations of motion is unchanged. Thus the Lagrangian is defined only to within an additive total time derivative of any function of coordinates and time.

§3. Galileo's relativity principle

In order to describe natural processes, it is necessary to choose a *frame of reference*. This is a coordinate system which serves to indicate the position of particles in space, together with a clock fixed to the system and serving to indicate time. The laws of nature, including the laws of motion, are in general different in form for different frames of reference. When an arbitrary frame of reference is chosen, it may happen that the laws governing even very simple phenomena become very complex. The problem naturally arises of finding a frame of reference in which the laws of nature take their simplest form.

The simplest type of motion is that of a free particle, i.e. one which is not subject to any external interactions. There exist frames of reference in which *free motion* takes place with a velocity which is constant in magnitude and direction. These are called *inertial frames*, and the statement that they exist is the *law of inertia*.

The inertial property may also be formulated by stating that space is homogeneous and isotropic, and time homogeneous, relative to such a frame of reference. The homogeneity of space and time signifies that all positions of a free particle in space at all times are equivalent; the isotropy of space, that different directions in space are equivalent. These properties evidently imply that free motion of a particle in any direction in space is an unchanging motion.

If two frames of reference are in uniform relative motion in a straight line, and one is an inertial frame, the other frame is obviously also an inertial frame, since any free motion in it takes place with constant velocity. Thus there is an infinity of inertial frames moving with constant relative velocities.

It is found, in fact, that different inertial frames are equivalent not only as regards the properties of free motion. Experiment shows that the *relativity principle* is valid: all the laws of nature are the same in every inertial frame. That is to say, the equations which express the laws of nature are invariant with respect to transformations of the coordinates and time from one inertial frame to another. This means that the equations have the same form when written in terms of the coordinates and time in different inertial frames.

Classical mechanics† or *Newtonian mechanics* is founded not only on the relativity principle but also on the hypothesis that *time is absolute*, i.e. that time runs in the same manner in every inertial frame. When combined with this hypothesis, the relativity principle is called *Galileo's relativity principle.*

The coordinates $\mathbf{r}$ and $\mathbf{r}'$ of a given point in two different inertial frames of reference K and K', of which the latter moves relative to the former with velocity $\mathbf{V}$, are related by

$$\mathbf{r} = \mathbf{r}' + \mathbf{V}t. \tag{3.1}$$

Here t, the time, is the same in the two frames:

$$t = t'. \tag{3.2}$$

Differentiation of equation (3.1) with respect to time gives the customary *law of composition of velocities*:

$$\mathbf{v} = \mathbf{v}' + \mathbf{V}. \tag{3.3}$$

Formulae (3.1) and (3.2) are called a *Galileo transformation*. Galileo's relativity principle states that the laws of nature are invariant under this transformation.

The foregoing discussion indicates quite clearly the exceptional properties of inertial frames, and consequently these frames should usually be employed in the investigation of mechanical phenomena. Unless otherwise stated, it will be implied henceforward that an inertial frame is used.

The complete physical equivalence of all inertial frames shows also that there is no "absolute" frame of reference which should be preferred to all other frames.

---

† As distinct from *relativistic mechanics* or *Einsteinian mechanics*, which will be discussed in Chapters 8 and 9.

§4. The Lagrangian for a free particle

Let us now go on to determine the form of the Lagrangian, and consider first of all the simplest case, that of the free motion of a particle relative to an inertial frame of reference.

Because of the homogeneity of space and time, the Lagrangian of a free particle cannot depend explicitly on either the position vector $\mathbf{r}$ or the time t, i.e. L is a function of the velocity $\mathbf{v}$ only; because of the isotropy of space, the Lagrangian must also be independent of the direction of the vector $\mathbf{v}$, and is therefore a function only of its magnitude, i.e. of $\mathbf{v}^2 = v^2$:

$$L = L(v^2).$$

The form of this function is uniquely determined by Galileo's relativity principle, which shows that $L(v^2)$ must have the same form in every inertial frame of reference. When we change from one frame to another, the velocity of the particle is transformed in accordance with (3.3), and $L(v^2)$ thus becomes $L[(\mathbf{v}'+\mathbf{V})^2]$. The latter expression must therefore differ from $L(v'^2)$, if at all, by the total derivative of a function of coordinates and time; it has been shown at the end of §2 that such a derivative can always be omitted from the Lagrangian.

This condition is satisfied only by a function of the form

$$L = av^2.$$

The transformation $\mathbf{v} = \mathbf{v}'+\mathbf{V}$ gives

$$\begin{aligned} L(v^2) &= av^2 = a(\mathbf{v}'+\mathbf{V})^2 \\ &= av'^2+2a\mathbf{v}'\cdot\mathbf{V}+aV^2, \end{aligned}$$

or, since $\mathbf{v}' = \mathrm{d}\mathbf{r}'/\mathrm{d}t$,

$$L(v^2) = L(v'^2)+\frac{\mathrm{d}}{\mathrm{d}t}(2a\mathbf{r}'\cdot\mathbf{V}+aV^2t).$$

The additional terms do in fact form a total derivative and may be omitted.

The constant a is usually written as $\frac{1}{2}m$, and the Lagrangian of a freely moving particle thus becomes

$$L = \tfrac{1}{2}mv^2. \tag{4.1}$$

The quantity m is called the *mass* of the particle. The additive property of the Lagrangian shows that for a system of particles which do not interact we have†

$$L = \sum \tfrac{1}{2}m_a v_a^2. \tag{4.2}$$

It should be emphasised that the above definition of mass becomes meaningful only when the additive property is taken into account. As has been mentioned in §2, the Lagrangian can always be multiplied by any constant without affecting the equations of motion. As regards the function (4.2), such multiplication amounts to a change in the unit of mass; the ratios of the masses of different particles remain unchanged thereby, and it is only these ratios which are physically meaningful.

It is easy to see that the mass of a particle cannot be negative. For, according to the principle of least action, the integral

$$S = \int_1^2 \tfrac{1}{2}mv^2 \, \mathrm{d}t$$

has a minimum for the actual motion of the particle in space from point 1 to point 2. If the mass were negative, the action integral would take arbitrarily large negative values for a motion in which the particle rapidly left point 1 and rapidly approached point 2, and there would be no minimum.

It is useful to notice that

$$v^2 = (\mathrm{d}l/\mathrm{d}t)^2 = (\mathrm{d}l)^2/(\mathrm{d}t)^2. \tag{4.3}$$

Hence, to obtain the Lagrangian, it is sufficient to find the square of the element of arc $\mathrm{d}l$ in a given system of coordinates. In Cartesian coordinates, for example, $\mathrm{d}l^2 = \mathrm{d}x^2 + \mathrm{d}y^2 + \mathrm{d}z^2$, and so

$$L = \tfrac{1}{2}m(\dot{x}^2 + \dot{y}^2 + \dot{z}^2). \tag{4.4}$$

---

† We shall use the suffixes $a, b, c, \ldots$ to distinguish the various particles, and $i, k, l, \ldots$ to distinguish the coordinates.

In cylindrical coordinates $dl^2 = dr^2+r^2\,d\phi^2+dz^2$, whence

$$L = \tfrac{1}{2}m(\dot{r}^2+r^2\dot{\phi}^2+\dot{z}^2). \tag{4.5}$$

In spherical coordinates $dl^2 = dr^2+r^2\,d\theta^2+r^2\sin^2\theta\,d\phi^2$, and

$$L = \tfrac{1}{2}m(\dot{r}^2+r^2\dot{\theta}^2+r^2\dot{\phi}^2\sin^2\theta). \tag{4.6}$$

§5. The Lagrangian for a system of particles

Let us now consider a system of particles which interact with one another but with no other bodies. This is called a *closed system*. It is found that the interaction between the particles can be described by adding to the Lagrangian (4.2) for non-interacting particles a certain function of the coordinates, which depends on the nature of the interaction.† Denoting this function by $-U$, we have

$$L = \sum \tfrac{1}{2}m_a v_a^2 - U(\mathbf{r}_1, \mathbf{r}_2, \ldots), \tag{5.1}$$

where $\mathbf{r}_a$ is the position of the ath particle. This is the general form of the Lagrangian for a closed system. The sum $T = \Sigma\tfrac{1}{2}m_a v_a^2$ is called the *kinetic energy*, and U the *potential energy*, of the system. The significance of these names is explained in §6.

The fact that the potential energy depends only on the positions of the particles at a given instant shows that a change in the position of any particle instantaneously affects all the other particles. We may say that the interactions are instantaneously propagated. The necessity for interactions in classical mechanics to be of this type is closely related to the premises upon which the subject is based, namely the absolute nature of time and Galileo's relativity principle. If the propagation of interactions were not instantaneous, but took place with a finite velocity, then that velocity would be different in different frames of reference in relative motion, since the absoluteness of time necessarily implies that the ordinary law of composition of velocities is applicable to all phenomena. The laws of motion for interacting

† This statement is valid only in classical mechanics.

bodies would then be different in different inertial frames, a result which would contradict the relativity principle.

In §3 only the homogeneity of time has been spoken of. The form of the Lagrangian (5.1) shows that time in mechanics is both homogeneous and isotropic, i.e. its properties are the same in both directions. For, if t is replaced by $-t$ (*time reversal*), the Lagrangian is unchanged, and therefore so are the equations of motion. In other words, if a given motion is possible in a system, then so is the reverse motion (that is, the motion in which the system passes through the same states in the reverse order). In this sense all motions which obey the laws of classical mechanics are reversible.

Knowing the Lagrangian, we can derive the equations of motion†:

$$\frac{\mathrm{d}}{\mathrm{d}t}\frac{\partial L}{\partial \mathbf{v}_a} = \frac{\partial L}{\partial \mathbf{r}_a}. \tag{5.2}$$

Substitution of (5.1) gives

$$m_a \,\mathrm{d}\mathbf{v}_a/\mathrm{d}t = -\partial U/\partial \mathbf{r}_a. \tag{5.3}$$

In this form the equations of motion are called *Newton's equations* and form the basis of the mechanics of a system of interacting particles. The vector

$$\mathbf{F}_a = -\partial U/\partial \mathbf{r}_a \tag{5.4}$$

which appears on the right-hand side of equation (5.3) is called the *force* on the ath particle. Like U, it depends only on the coordinates of the particles, and not on their velocities. The equation (5.3) therefore shows that the acceleration vectors of the particles are likewise functions of their coordinates only.

The potential energy is defined only to within an additive constant, which has no effect on the equations of motion. This is a particular case of the non-uniqueness of the Lagrangian discussed at the end of §2. The most natural and most usual way of choosing this constant

† The derivative of a scalar quantity with respect to a vector is defined as the vector whose components are equal to the derivatives of the scalar with respect to the corresponding components of the vector.

is such that the potential energy tends to zero as the distances between the particles tend to infinity.

If we use, to describe the motion, arbitrary generalised coordinates q_i instead of Cartesian coordinates, the following transformation is needed to obtain the new Lagrangian:

$$x_a = f_a(q_1, q_2, \ldots, q_s), \quad \dot{x}_a = \sum_k \frac{\partial f_a}{\partial q_k} \dot{q}_k.$$

Substituting these expressions in the function $L = \frac{1}{2}\sum m_a(\dot{x}_a^2+\dot{y}_a^2+\dot{z}_a^2)-U$, we obtain the required Lagrangian in the form

$$L = \tfrac{1}{2} \sum_{i,k} a_{ik}(q)\dot{q}_i\dot{q}_k - U(q), \tag{5.5}$$

where the a_{ik} are functions of the coordinates only. The kinetic energy in generalised coordinates is still a quadratic function of the velocities, but it may depend on the coordinates also.

Hitherto we have spoken only of closed systems. Let us now consider a system A which is not closed and interacts with another system B executing a given motion. In such a case we say that the system A moves in a given external field (due to the system B). Since the equations of motion are obtained from the principle of least action by independently varying each of the coordinates (i.e. by proceeding as if the remainder were given quantities), we can find the Lagrangian L_A of the system A by using the Lagrangian L of the whole system $A+B$ and replacing the coordinates q_B therein by given functions of time.

Assuming that the system $A+B$ is closed, we have $L = T_A(q_A, \dot{q}_A)+T_B(q_B, \dot{q}_B)-U(q_A, q_B)$, where the first two terms are the kinetic energies of the systems A and B and the third term is their combined potential energy. Substituting for q_B the given functions of time and omitting the term $T[q_B(t), \dot{q}_B(t)]$ which depends on time only, and is therefore the total time derivative of a function of time, we obtain $L_A = T_A(q_A, \dot{q}_A)-U[q_A, q_B(t)]$. Thus the motion of a system in an external field is described by a Lagrangian of the usual type, the only difference being that the potential energy may depend explicitly on time.

For example, when a single particle moves in an external field, the general form of the Lagrangian is

$$L = \tfrac{1}{2}mv^2 - U(\mathbf{r}, t), \tag{5.6}$$

and the equation of motion is

$$m\dot{\mathbf{v}} = -\partial U/\partial \mathbf{r}. \tag{5.7}$$

A field such that the same force **F** acts on a particle at any point in the field is said to be *uniform*. The potential energy in such a field is evidently

$$U = -\mathbf{F}\cdot\mathbf{r}. \tag{5.8}$$

To conclude this section, we may make the following remarks concerning the application of Lagrange's equations to various problems. It is often necessary to deal with mechanical systems in which the interaction between different bodies (or particles) takes the form of *constraints*, i.e. restrictions on their relative position. In practice, such constraints are effected by means of rods, strings, hinges and so on. This introduces a new factor into the problem, in that the motion of the bodies results in friction at their points of contact, and the problem in general ceases to be one of pure mechanics (see §20). In many cases, however, the friction in the system is so slight that its effect on the motion is entirely negligible. If the masses of the constraining elements of the system are also negligible, the effect of the constraints is simply to reduce the number of degrees of freedom s of the system to a value less than $3N$. To determine the motion of the system, the Lagrangian (5.5) can again be used, with a set of independent generalised coordinates equal in number to the actual degrees of freedom.

PROBLEMS

Find the Lagrangian for each of the following systems when placed in a uniform gravitational field (acceleration g).

PROBLEM 1. A coplanar double pendulum (Fig. 1).

SOLUTION. We take as coordinates the angles ϕ_1 and ϕ_2 which the strings l_1 and l_2 make with the vertical. Then we have, for the particle m_1, $T_1 = \tfrac{1}{2}m_1 l_1^2 \dot{\phi}_1^2$, $U = -m_1 g l_1 \cos\phi_1$. In order to find the kinetic energy of the second particle, we express its Cartesian coordinates x_2, y_2 (with the origin at the point of support and

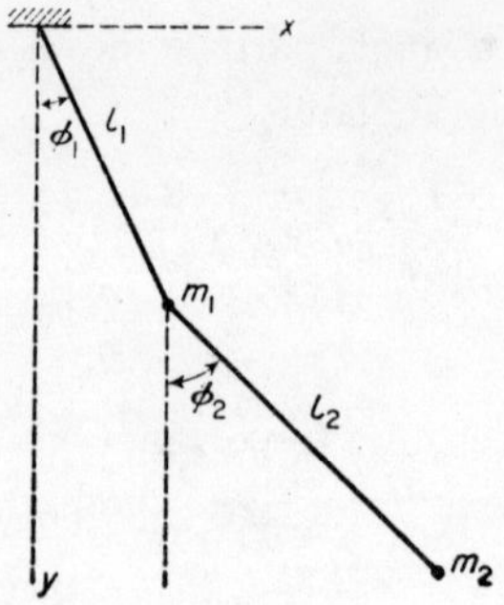

FIG. 1

the y axis vertically downwards) in terms of the angles ϕ_1 and ϕ_2: $x_2 = l_1 \sin\phi_1 + l_2 \sin\phi_2$, $y_2 = l_1 \cos\phi_1 + l_2 \cos\phi_2$. Then we find

$$T_2 = \tfrac{1}{2}m_2(\dot{x}_2^2+\dot{y}_2^2)$$
$$= \tfrac{1}{2}m_2[l_1^2\dot{\phi}_1^2+l_2^2\dot{\phi}_2^2+2l_1l_2\cos(\phi_1-\phi_2)\dot{\phi}_1\dot{\phi}_2].$$

Finally

$$L = \tfrac{1}{2}(m_1+m_2)l_1^2\dot{\phi}_1^2+\tfrac{1}{2}m_2l_2^2\dot{\phi}_2^2+m_2l_1l_2\dot{\phi}_1\dot{\phi}_2\cos(\phi_1-\phi_2)+(m_1+m_2)gl_1\cos\phi_1+ +m_2gl_2\cos\phi_2.$$

PROBLEM 2. A simple pendulum whose point of support oscillates vertically according to the law $a \cos \gamma t$ (Fig. 2).

SOLUTION. The coordinates of m are

$$x = l\sin\phi, \qquad y = l\cos\phi + a\cos\gamma t.$$

The Lagrangian is

$$L = \tfrac{1}{2}ml^2\dot{\phi}^2 + mla\gamma^2\cos\gamma t\cos\phi + mgl\cos\phi;$$

here terms depending only on time have been omitted, together with the total time derivative of $mla\gamma^2 \sin\gamma t \cos\phi$.

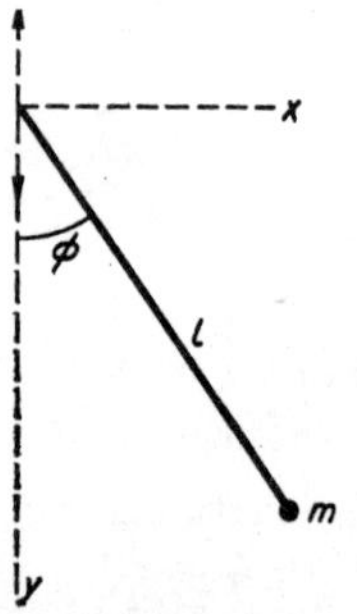

FIG. 2

CHAPTER 2

CONSERVATION LAWS

§6. Energy

During the motion of a mechanical system, the $2s$ quantities q_i and $\dot{q}_i$ ($i = 1, 2, \ldots, s$) which specify the state of the system vary with time. There exist, however, functions of these quantities whose values remain constant during the motion, and depend only on the initial conditions. Such functions are called *integrals of the motion.*

The number of independent integrals of the motion for a closed mechanical system with s degrees of freedom is $2s-1$. This is evident from the following simple arguments. The general solution of the equations of motion contains $2s$ arbitrary constants (see the discussion following equation (2.6)). Since the equations of motion for a closed system do not involve the time explicitly, the choice of the origin of time is entirely arbitrary, and one of the arbitrary constants in the solution of the equations can always be taken as an additive constant t_0 in the time. Eliminating $t+t_0$ from the $2s$ functions $q_i = q_i(t+t_0, C_1, C_2, \ldots, C_{2s-1})$, $\dot{q}_i = \dot{q}_i(t+t_0, C_1, C_2, \ldots, C_{2s-1})$, we can express the $2s-1$ arbitrary constants $C_1, C_2, \ldots, C_{2s-1}$ as functions of q and $\dot{q}$, and these functions will be integrals of the motion.

Not all integrals of the motion, however, are of equal importance in mechanics. There are some whose constancy is of profound significance, deriving from the fundamental homogeneity and isotropy of space and time. The quantities represented by such integrals of the motion are said to be *conserved,* and have an important common property of being additive: their values for a system composed of several parts which do not interact are equal to the sums of their values for the individual parts.

It is to this additivity that the quantities concerned owe their especial importance in mechanics. Let us suppose, for example, that two bodies interact only during a certain interval of time. Since each of the additive integrals of the whole system is, both before and after the interaction, equal to the sum of its values for the two bodies separately, the conservation laws for these quantities immediately make possible various conclusions regarding the state of the bodies after the interaction, if their states before the interaction are known.

Let us consider first the conservation law resulting from the *homogeneity of time*. By virtue of this homogeneity, the Lagrangian of a closed system does not depend explicitly on time. The total time derivative of the Lagrangian can therefore be written

$$\frac{\mathrm{d}L}{\mathrm{d}t} = \sum_i \frac{\partial L}{\partial q_i} \dot{q}_i + \sum_i \frac{\partial L}{\partial \dot{q}_i} \ddot{q}_i .$$

If L depended explicitly on time, a term $\partial L/\partial t$ would have to be added on the right-hand side. Replacing $\partial L/\partial q_i$, in accordance with Lagrange's equations, by $(\mathrm{d}/\mathrm{d}t)\, \partial L/\partial \dot{q}_i$, we obtain

$$\frac{\mathrm{d}L}{\mathrm{d}t} = \sum_i \dot{q}_i \frac{\mathrm{d}}{\mathrm{d}t}\left(\frac{\partial L}{\partial \dot{q}_i}\right) + \sum_i \frac{\partial L}{\partial \dot{q}_i} \ddot{q}_i$$

$$= \sum_i \frac{\mathrm{d}}{\mathrm{d}t}\left(\dot{q}_i \frac{\partial L}{\partial \dot{q}_i}\right)$$

or

$$\frac{\mathrm{d}}{\mathrm{d}t}\left(\sum_i \dot{q}_i \frac{\partial L}{\partial \dot{q}_i} - L\right) = 0.$$

Hence we see that the quantity

$$E \equiv \sum_i \dot{q}_i \frac{\partial L}{\partial \dot{q}_i} - L \tag{6.1}$$

remains constant during the motion of a closed system, i.e. it is an integral of the motion; it is called the *energy* of the system. The additivity of the energy follows immediately from that of the Lagrangian, since (6.1) shows that it is a linear function of the latter.

The law of conservation of energy is valid not only for closed systems, but also for those in a constant external field (i.e. one independent of time): the only property of the Lagrangian used in the above derivation, namely that it does not involve the time explicitly, is still valid. Mechanical systems whose energy is conserved are sometimes called *conservative* systems.

As we have seen in §5, the Lagrangian of a closed system (or one in a constant field) is of the form $L = T(q, \dot{q}) - U(q)$, where T is a quadratic function of the velocities. Using Euler's theorem on homogeneous functions, we have

$$\sum_i \dot{q}_i \frac{\partial L}{\partial \dot{q}_i} = \sum_i \dot{q}_i \frac{\partial T}{\partial \dot{q}_i} = 2T.$$

Substituting this in (6.1) gives

$$E = T(q, \dot{q}) + U(q); \qquad (6.2)$$

in Cartesian coordinates,

$$E = \sum_a \tfrac{1}{2} m_a v_a^2 + U(\mathbf{r}_1, \mathbf{r}_2, \ldots). \qquad (6.3)$$

Thus the energy of the system can be written as the sum of two quite different terms: the kinetic energy, which depends on the velocities, and the potential energy, which depends only on the coordinates of the particles.

§7. Momentum

A second conservation law follows from the *homogeneity of space*. By virtue of this homogeneity, the mechanical properties of a closed system are unchanged by any parallel displacement of the entire system in space. Let us therefore consider an infinitesimal displacement $\boldsymbol{\varepsilon}$, and obtain the condition for the Lagrangian to remain unchanged.

A parallel displacement is a transformation in which every particle in the system is moved by the same amount, the position vector $\mathbf{r}$ becoming $\mathbf{r} + \boldsymbol{\varepsilon}$. The change in L resulting from an infinitesimal change in the coordinates, the velocities of the particles remaining fixed, is

$$\delta L = \sum_a \frac{\partial L}{\partial \mathbf{r}_a} \cdot \delta \mathbf{r}_a = \boldsymbol{\varepsilon} \cdot \sum_a \frac{\partial L}{\partial \mathbf{r}_a},$$

where the summation is over the particles in the system. Since $\boldsymbol{\varepsilon}$ is arbitrary, the condition $\delta L = 0$ is equivalent to

$$\sum_a \partial L / \partial \mathbf{r}_a = 0. \tag{7.1}$$

From Lagrange's equations (5.2) we therefore have

$$\sum_a \frac{\mathrm{d}}{\mathrm{d}t} \frac{\partial L}{\partial \mathbf{v}_a} = \frac{\mathrm{d}}{\mathrm{d}t} \sum_a \frac{\partial L}{\partial \mathbf{v}_a} = 0.$$

Thus we conclude that, in a closed mechanical system, the vector

$$\mathbf{P} \equiv \sum_a \partial L / \partial \mathbf{v}_a \tag{7.2}$$

remains constant during the motion; it is called the *momentum* of the system. Differentiating the Lagrangian (5.1), we find that the momentum is given in terms of the velocities of the particles by

$$\mathbf{P} = \sum_a m_a \mathbf{v}_a. \tag{7.3}$$

The additivity of the momentum is evident. Moreover, unlike the energy, the momentum of the system is equal to the sum of its values $\mathbf{p}_a = m_a \mathbf{v}_a$ for the individual particles, whether or not the interaction between them can be neglected.

The three components of the momentum vector are all conserved only in the absence of an external field. The individual components may be conserved even in the presence of a field, however, if the potential energy in the field does not depend on all the Cartesian coordinates. The mechanical properties of the system are evidently unchanged by a displacement along the axis of a coordinate which does not appear in the potential energy, and so the corresponding component of the momentum is conserved. For example, in a uniform field in the z direction, the x and y components of momentum are conserved.

The equation (7.1) has a simple physical meaning. The derivative $\partial L/\partial \mathbf{r}_a = -\partial U/\partial \mathbf{r}_a$ is the force $\mathbf{F}_a$ acting on the ath particle. Thus equation (7.1) signifies that the sum of the forces on all the particles in a closed system is zero:

$$\sum_a \mathbf{F}_a = 0. \tag{7.4}$$

In particular, for a system of only two particles, $\mathbf{F}_1 + \mathbf{F}_2 = 0$: the force exerted by the first particle on the second is equal in magnitude, and opposite in direction, to that exerted by the second particle on the first. This is the equality of action and reaction (*Newton's third law*).

If the motion is described by generalised coordinates q_i, the derivatives of the Lagrangian with respect to the generalised velocities

$$p_i = \partial L/\partial \dot{q}_i \tag{7.5}$$

are called *generalised momenta*, and its derivatives with respect to the generalised coordinates

$$F_i = \partial L/\partial q_i \tag{7.6}$$

are called *generalised forces*. In this notation, Lagrange's equations are

$$\dot{p}_i = F_i. \tag{7.7}$$

In Cartesian coordinates the generalised momenta are the components of the vectors $\mathbf{p}_a$. In general, however, the p_i are linear homogeneous functions of the generalised velocities $\dot{q}_i$, and do not reduce to products of mass and velocity.

§8. Centre of mass

The momentum of a closed mechanical system has different values in different (inertial) frames of reference. If a frame K' moves with velocity $\mathbf{V}$ relative to another frame K, then the velocities $\mathbf{v}'_a$ and $\mathbf{v}_a$ of the particles relative to the two frames are such that $\mathbf{v}_a = \mathbf{v}'_a + \mathbf{V}$.

The momenta $\mathbf{P}$ and $\mathbf{P}'$ in the two frames are therefore related by

$$\mathbf{P} = \sum_a m_a \mathbf{v}_a = \sum_a m_a \mathbf{v}'_a + \mathbf{V} \sum_a m_a,$$

or

$$\mathbf{P} = \mathbf{P}' + \mathbf{V} \sum_a m_a. \tag{8.1}$$

In particular, there is always a frame of reference K' in which the total momentum is zero. Putting $\mathbf{P}' = 0$ in (8.1), we find the velocity of this frame:

$$\mathbf{V} = \mathbf{P}/\sum m_a = \sum m_a \mathbf{v}_a / \sum m_a. \tag{8.2}$$

If the total momentum of a mechanical system in a given frame of reference is zero, it is said to be *at rest* relative to that frame. This is a natural generalisation of the term as applied to a particle. Similarly, the velocity $\mathbf{V}$ given by (8.2) is the velocity of the "motion as a whole" of a mechanical system whose momentum is not zero. Thus we see that the law of conservation of momentum makes possible a natural definition of rest and velocity, as applied to a mechanical system as a whole.

Formula (8.2) shows that the relation between the momentum $\mathbf{P}$ and the velocity $\mathbf{V}$ of the system is the same as that between the momentum and velocity of a single particle of mass $\mu = \Sigma m_a$, the sum of the masses of the particles in the system. This result can be regarded as expressing the *additivity of mass.*

The right-hand side of formula (8.2) can be written as the total time derivative of the expression

$$\mathbf{R} \equiv \sum m_a \mathbf{r}_a / \sum m_a. \tag{8.3}$$

We can say that the velocity of the system as a whole is the rate of motion in space of the point whose position is (8.3). This point is called the *centre of mass* of the system.

The law of conservation of momentum for a closed system can be formulated as stating that the centre of mass of the system moves uniformly in a straight line. In this form it generalises the law of inertia derived in §3 for a single free particle, whose "centre of mass" coincides with the particle itself.

In considering the mechanical properties of a closed system it is natural to use a frame of reference in which the centre of mass is at rest. This eliminates a uniform rectilinear motion of the system as a whole.

The energy of a mechanical system which is at rest as a whole is usually called its *internal energy* E_i. This includes the kinetic energy of the relative motion of the particles in the system and the potential energy of their interaction. The total energy of a system moving as a whole with velocity V can be written

$$E = \tfrac{1}{2}\mu V^2 + E_i. \tag{8.4}$$

Although this formula is fairly obvious, we may give a direct proof of it. The energies E and E' of a mechanical system in two frames of reference K and K' are related by

$$\begin{aligned} E &= \tfrac{1}{2}\sum_a m_a v_a^2 + U \\ &= \tfrac{1}{2}\sum_a m_a(\mathbf{v}'_a + \mathbf{V})^2 + U \\ &= \tfrac{1}{2}\mu V^2 + \mathbf{V}\cdot\sum_a m_a \mathbf{v}'_a + \tfrac{1}{2}\sum_a m_a v_a'^2 + U \end{aligned}$$

or

$$E = E' + \mathbf{V}\cdot\mathbf{P}' + \tfrac{1}{2}\mu V^2. \tag{8.5}$$

This formula gives the law of transformation of energy from one frame to another, corresponding to formula (8.1) for momentum. If the centre of mass is at rest in K', then $\mathbf{P}' = 0$, $E' = E_i$, and we have (8.4).

§9. Angular momentum

Let us now derive the conservation law which follows from the *isotropy of space*. This isotropy means that the mechanical properties of a closed system do not vary when it is rotated as a whole in any manner in space. Let us therefore consider an infinitesimal rotation of the system, and obtain the condition for the Lagrangian to remain unchanged.

We shall use the vector $\delta\boldsymbol{\phi}$ of the infinitesimal rotation, whose magnitude is the angle of rotation $\delta\phi$, and whose direction is that of the axis of rotation (the direction of rotation being that of a right-handed screw driven along $\delta\boldsymbol{\phi}$).

Let us find, first of all, the resulting increment in the position vector from an origin on the axis to any particle in the system undergoing rotation. The linear displacement of the end of the position vector

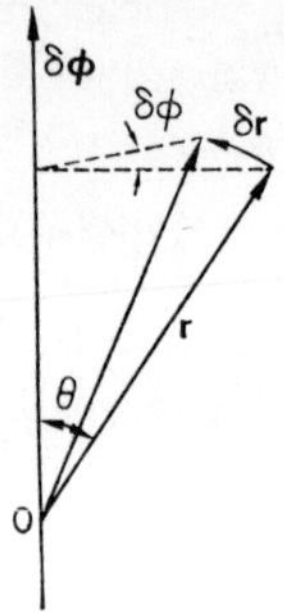

FIG. 3

is related to the angle by $|\delta\mathbf{r}| = r \sin\theta\, \delta\phi$ (Fig. 3). The direction of $\delta\mathbf{r}$ is perpendicular to the plane of $\mathbf{r}$ and $\delta\boldsymbol{\phi}$. Hence it is clear that

$$\delta\mathbf{r} = \delta\boldsymbol{\phi}\times\mathbf{r}. \tag{9.1}$$

When the system is rotated, not only the position vectors but also the velocities of the particles change direction, and all vectors are transformed in the same manner. The velocity increment relative to a fixed system of coordinates is therefore

$$\delta\mathbf{v} = \delta\boldsymbol{\phi}\times\mathbf{v}. \tag{9.2}$$

If these expressions are substituted in the condition that the Lagrangian is unchanged by the rotation:

$$\delta L = \sum_a \left(\frac{\partial L}{\partial \mathbf{r}_a}\cdot\delta\mathbf{r}_a + \frac{\partial L}{\partial \mathbf{v}_a}\cdot\delta\mathbf{v}_a\right) = 0$$

and the derivative $\partial L/\partial\mathbf{v}_a$ replaced by $\mathbf{p}_a$, and $\partial L/\partial\mathbf{r}_a$ by $\dot{\mathbf{p}}_a$, the result is

$$\sum_a (\dot{\mathbf{p}}_a\cdot\delta\boldsymbol{\phi}\times\mathbf{r}_a + \mathbf{p}_a\cdot\delta\boldsymbol{\phi}\times\mathbf{v}_a) = 0$$

or, permuting the factors and taking $\delta\boldsymbol{\phi}$ outside the sum,

$$\delta\boldsymbol{\phi}\cdot\sum_a(\mathbf{r}_a\times\dot{\mathbf{p}}_a+\mathbf{v}_a\times\mathbf{p}_a) = \delta\boldsymbol{\phi}\cdot\frac{\mathrm{d}}{\mathrm{d}t}\sum_a\mathbf{r}_a\times\mathbf{p}_a = 0.$$

Since $\delta\boldsymbol{\phi}$ is arbitrary, it follows that $(\mathrm{d}/\mathrm{d}t)\ \Sigma\mathbf{r}_a\times\mathbf{p}_a = 0$, and we conclude that the vector

$$\mathbf{M}\equiv\sum_a\mathbf{r}_a\times\mathbf{p}_a, \tag{9.3}$$

called the *angular momentum* or *moment of momentum* of the system, is conserved in the motion of a closed system. Like the linear momentum, it is additive, whether or not the particles in the system interact.

There are no other additive integrals of the motion. Thus every closed system has seven such integrals: energy, three components of momentum, and three components of angular momentum.

Since the definition of angular momentum involves the positions of the particles, its value depends in general on the choice of origin. The positions $\mathbf{r}_a$ and $\mathbf{r}'_a$ of a given point relative to origins at a distance $\mathbf{a}$ apart are related by $\mathbf{r}_a = \mathbf{r}'_a+\mathbf{a}$. Hence

$$\begin{aligned}\mathbf{M} &= \sum_a\mathbf{r}_a\times\mathbf{p}_a\\ &= \sum_a\mathbf{r}'_a\times\mathbf{p}_a+\mathbf{a}\times\sum_a\mathbf{p}_a\end{aligned}$$

or

$$\mathbf{M} = \mathbf{M}'+\mathbf{a}\times\mathbf{P}. \tag{9.4}$$

It is seen from this formula that the angular momentum depends on the choice of origin except when the system is at rest as a whole (i.e. $\mathbf{P} = 0$). This indeterminacy, of course, does not affect the law of conservation of angular momentum, since momentum is also conserved in a closed system.

We may also derive a relation between the angular momenta in two inertial frames of reference K and K', of which the latter moves with velocity $\mathbf{V}$ relative to the former. We shall suppose that the origins in the frames K and K' coincide at a given instant. Then the position vectors of the particles are the same in the two frames, while their velocities are related by $\mathbf{v}_a = \mathbf{v}'_a + \mathbf{V}$. Hence we have

$$\mathbf{M} = \sum_a m_a \mathbf{r}_a \times \mathbf{v}_a = \sum_a m_a \mathbf{r}_a \times \mathbf{v}'_a + \sum_a m_a \mathbf{r}_a \times \mathbf{V}.$$

The first sum on the right-hand side is the angular momentum $\mathbf{M}'$ in the frame K'; using in the second sum the position of the centre of mass (8.3), we obtain

$$\mathbf{M} = \mathbf{M}' + \mu \mathbf{R} \times \mathbf{V}. \tag{9.5}$$

This formula gives the law of transformation of angular momentum from one frame to another, corresponding to formula (8.1) for momentum and (8.5) for energy.

If the frame K' is that in which the system considered is at rest as a whole, then $\mathbf{V}$ is the velocity of its centre of mass, $\mu\mathbf{V}$ its total momentum $\mathbf{P}$ relative to K, and

$$\mathbf{M} = \mathbf{M}' + \mathbf{R} \times \mathbf{P}. \tag{9.6}$$

In other words, the angular momentum $\mathbf{M}$ of a mechanical system consists of its "intrinsic angular momentum" in a frame in which it is at rest, and the angular momentum $\mathbf{R} \times \mathbf{P}$ due to its motion as a whole.

Although the law of conservation of all three components of angular momentum (relative to an arbitrary origin) is valid only for a closed system, the law of conservation may hold in a more restricted form even for a system in an external field. It is evident from the above derivation that the component of angular momentum along an axis about which the field is symmetrical is always conserved, for the mechanical properties of the system are unaltered by any rotation about that axis. Here the angular momentum must, of course, be defined relative to an origin lying on the axis.

The most important such case is that of a *centrally symmetric field* or *central field*, i.e. one in which the potential energy depends only on the distance from some particular point (the *centre*). It is evident that the component of angular momentum along any axis passing through the centre is conserved in motion in such a field. In other words, the angular momentum **M** is conserved provided that it is defined with respect to the centre of the field.

Another example is that of a field homogeneous in the z direction; in such a field, the component M_z of the angular momentum is conserved, whichever point is taken as the origin.

The component of angular momentum along any axis (say the z axis) can be found by differentiation of the Lagrangian:

$$M_z = \sum_a \frac{\partial L}{\partial \dot{\phi}_a}, \tag{9.7}$$

where the coordinate ϕ is the angle of rotation about the z axis. This is evident from the above proof of the law of conservation of angular momentum, but can also be proved directly. In cylindrical coordinates r, ϕ, z we have (substituting $x_a = r_a \cos \phi_a$, $y_a = r_a \sin \phi_a$)

$$\begin{aligned} M_z &= \sum_a m_a(x_a \dot{y}_a - y_a \dot{x}_a) \\ &= \sum_a m_a r_a^2 \dot{\phi}_a. \end{aligned} \tag{9.8}$$

The Lagrangian is, in terms of these coordinates,

$$L = \tfrac{1}{2} \sum_a m_a(\dot{r}_a^2 + r_a^2 \dot{\phi}_a^2 + \dot{z}_a^2) - U,$$

and substitution of this in (9.7) gives (9.8).

PROBLEM

Which components of momentum **P** and angular momentum **M** are conserved in motion in the following fields?

(a) the field of an infinite homogeneous plane, (b) that of an infinite homogeneous cylinder, (c) that of an infinite homogeneous prism, (d) that of two points, (e) that of an infinite homogeneous half-plane, (f) that of a homogeneous cone.

SOLUTION. (a) P_x, P_y, M_z (if the plane is the xy plane), (b) M_z, P_z (if the axis of the cylinder is the z axis), (c) P_z (if the edges of the prism are parallel to the z axis), (d) M_z (if the line joining the points is the z axis), (e) P_y (if the edge of the half-plane is the y axis), (f) M_z (if the axis of the cone is the z axis).

CHAPTER 3

INTEGRATION OF THE EQUATIONS OF MOTION

§10. Motion in one dimension

The motion of a system having one degree of freedom is said to take place *in one dimension*. The most general form of the Lagrangian of such a system in fixed external conditions is

$$L = \tfrac{1}{2}a(q)\dot{q}^2 - U(q), \tag{10.1}$$

where $a(q)$ is some function of the generalised coordinate q. In particular, if q is a Cartesian coordinate (x, say) then

$$L = \tfrac{1}{2}m\dot{x}^2 - U(x). \tag{10.2}$$

The equations of motion corresponding to these Lagrangians can be integrated in a general form. It is not even necessary to write down the equation of motion; we can start from the first integral of this equation, which gives the law of conservation of energy. For the Lagrangian (10.2) (e.g.) we have $\frac{1}{2}m\dot{x}^2 + U(x) = E$. This is a first-order differential equation, and can be integrated immediately. Since $\mathrm{d}x/\mathrm{d}t = \sqrt{\{2[E-U(x)]/m\}}$, it follows that

$$t = \sqrt{(\tfrac{1}{2}m)}\int\frac{\mathrm{d}x}{\sqrt{[E-U(x)]}} + \text{constant}. \tag{10.3}$$

The two arbitrary constants in the solution of the equation of motion are here represented by the total energy E and the constant of integration.

Since the kinetic energy is essentially positive, the total energy always exceeds the potential energy, i.e. the motion can take place only in those regions of space where $U(x) < E$. For example, let the function $U(x)$ be of the form shown in Fig. 4. If we draw in the figure a horizontal line corresponding to a given value of the total energy, we immediately find the possible regions of motion. In the example of Fig. 4, the motion can occur only in the range AB or in the range to the right of C.

The points at which the potential energy equals the total energy,

$$U(x) = E, \tag{10.4}$$

give the limits of the motion. They are *turning points*, since the velocity there is zero. If the region of the motion is bounded by two such points, then the motion takes place in a finite region of space, and is said to be *finite*. If the region of the motion is limited on only one

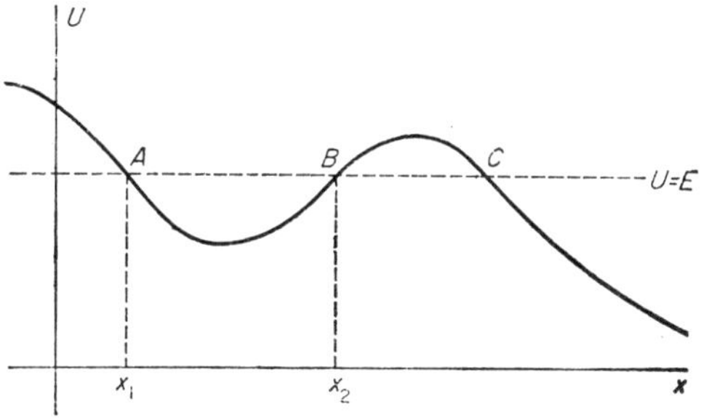

FIG. 4

side, or on neither, then the motion is *infinite* and the particle goes to infinity.

A finite motion in one dimension is oscillatory, the particle moving repeatedly back and forth between two points (in Fig. 4, in the *potential well* AB between the points x_1 and x_2). The period T of the oscillations, i.e. the time during which the particle passes from x_1 to x_2 and back, is twice the time from x_1 to x_2 (because of the reversibility property,

§5) or, by (10.3),

$$T(E) = \sqrt{(2m)} \int\limits_{x_1(E)}^{x_2(E)} \frac{dx}{\sqrt{[E-U(x)]}}, \tag{10.5}$$

where x_1 and x_2 are roots of equation (10.4) for the given value of E. This formula gives the period of the motion as a function of the total energy of the particle.

§11. The reduced mass

A complete general solution can be obtained for an extremely important problem, that of the motion of a system consisting of two interacting particles (the *two-body problem*).

As a first step towards the solution of this problem, we shall show how it can be considerably simplified by separating the motion of the system into the motion of the centre of mass and that of the particles relative to the centre of mass.

The potential energy of the interaction of two particles depends only on the distance between them, i.e. on the magnitude of the difference in their position vectors. The Lagrangian of such a system is therefore

$$L = \tfrac{1}{2}m_1\dot{\mathbf{r}}_1^2 + \tfrac{1}{2}m_2\dot{\mathbf{r}}_2^2 - U(|\mathbf{r}_1 - \mathbf{r}_2|). \tag{11.1}$$

Let $\mathbf{r} \equiv \mathbf{r}_1 - \mathbf{r}_2$ be the relative position vector, and let the origin be at the centre of mass, i.e. $m_1\mathbf{r}_1 + m_2\mathbf{r}_2 = 0$. These two equations give

$$\mathbf{r}_1 = m_2\mathbf{r}/(m_1+m_2), \qquad \mathbf{r}_2 = -m_1\mathbf{r}/(m_1+m_2). \tag{11.2}$$

Substitution in (11.1) gives

$$L = \tfrac{1}{2}m\dot{\mathbf{r}}^2 - U(r), \tag{11.3}$$

where

$$m = m_1m_2/(m_1+m_2) \tag{11.4}$$

is called the *reduced mass*. The function (11.3) is formally identical with the Lagrangian of a particle of mass m moving in an external field $U(r)$ which is symmetrical about a fixed origin.

Thus the problem of the motion of two interacting particles is equivalent to that of the motion of one particle in a given external field $U(r)$. From the solution $\mathbf{r} = \mathbf{r}(t)$ of this problem, the paths $\mathbf{r}_1 = \mathbf{r}_1(t)$ and $\mathbf{r}_2 = \mathbf{r}_2(t)$ of the two particles separately, relative to their common centre of mass, are obtained by means of formulae (11.2).

§12. Motion in a central field

On reducing the two-body problem to one of the motion of a single body, we arrive at the problem of determining the motion of a single particle in an external field such that its potential energy depends only on the distance r from some fixed point. This is called a *central* field. The force acting on the particle is $\mathbf{F} = -\partial U(r)/\partial \mathbf{r} = -(\mathrm{d}U/\mathrm{d}r)\mathbf{r}/r$; its magnitude is likewise a function of r only, and its direction is everywhere that of the position vector.

As has already been shown in §9, the angular momentum of any system relative to the centre of such a field is conserved. The angular momentum of a single particle is $\mathbf{M} = \mathbf{r}\times\mathbf{p}$. Since $\mathbf{M}$ is perpendicular to $\mathbf{r}$, the constancy of $\mathbf{M}$ shows that, throughout the motion, the position vector of the particle lies in the plane perpendicular to $\mathbf{M}$.

Thus the path of a particle in a central field lies in one plane. Using polar coordinates r, ϕ in that plane, we can write the Lagrangian as

$$L = \tfrac{1}{2}m(\dot{r}^2 + r^2\dot{\phi}^2) - U(r); \tag{12.1}$$

see (4.5). This function does not involve the coordinate ϕ explicitly. Any generalised coordinate q_i which does not appear explicitly in the Lagrangian is said to be *cyclic*. For such a coordinate we have, by Lagrange's equation, $(\mathrm{d}/\mathrm{d}t)\,\partial L/\partial \dot{q}_i = \partial L/\partial q_i = 0$, so that the corresponding generalised momentum $p_i = \partial L/\partial \dot{q}_i$ is an integral of the motion. This leads to a considerable simplification of the problem of integrating the equations of motion when there are cyclic coordinates.

In the present case, the generalised momentum $p_\phi = mr^2\dot{\phi}$ is the same as the angular momentum $M_z = M$ (see (9.8)), and we return to the known law of conservation of angular momentum:

$$M = mr^2\dot{\phi} = \text{constant}. \tag{12.2}$$

This law has a simple geometrical interpretation in the plane motion of a single particle in a central field. The expression $\frac{1}{2}\mathbf{r}\cdot r\mathrm{d}\phi$ is the area of the sector bounded by two neighbouring position vectors and an element of the path (Fig. 5). Calling this area df, we can write the angular momentum of the particle as

$$M = 2m\dot{f}, \tag{12.3}$$

where the derivative $\dot{f}$ is called the *sectorial velocity*. Hence the conservation of angular momentum implies the constancy of the sectorial velocity: in equal times the position vector of the particle sweeps out equal areas (*Kepler's second law*).†

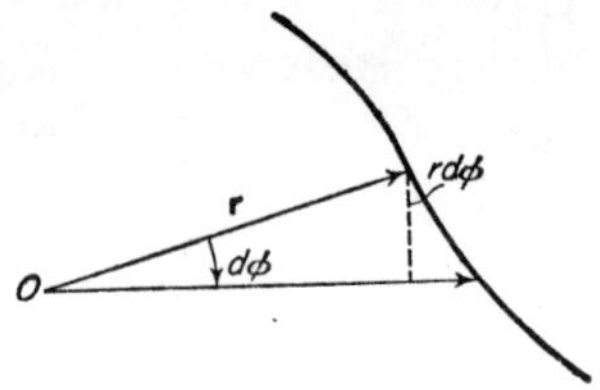

FIG. 5

The complete solution of the problem of the motion of a particle in a central field is most simply obtained by starting from the laws of conservation of energy and angular momentum, without writing out the equations of motion themselves. Expressing $\dot{\phi}$ in terms of M from (12.2) and substituting in the expression for the energy, we obtain

$$E = \tfrac{1}{2}m(\dot{r}^2+r^2\dot{\phi}^2)+U(r) = \tfrac{1}{2}m\dot{r}^2+\tfrac{1}{2}M^2/mr^2+U(r). \tag{12.4}$$

Hence

$$\dot{r} \equiv \frac{\mathrm{d}r}{\mathrm{d}t} = \sqrt{\left\{\frac{2}{m}[E-U(r)]-\frac{M^2}{m^2r^2}\right\}} \tag{12.5}$$

or, integrating,

$$t = \int \mathrm{d}r\bigg/\sqrt{\left\{\frac{2}{m}[E-U(r)]-\frac{M^2}{m^2r^2}\right\}}+\text{constant}. \tag{12.6}$$

† The law of conservation of angular momentum for a particle moving in a central field is sometimes called the *area integral*.

Writing (12.2) as $d\phi = M\, dt/mr^2$, substituting dt from (12.5) and integrating, we find

$$\phi = \int \frac{M\, dr/r^2}{\sqrt{\{2m[E-U(r)]-M^2/r^2\}}} + \text{constant}. \tag{12.7}$$

Formulae (12.6) and (12.7) give the general solution of the problem. The latter formula gives the relation between r and ϕ, i.e. the equation of the path. Formula (12.6) gives the distance r from the centre as an implicit function of time. The angle ϕ, it should be noted, always varies monotonically with time, since (12.2) shows that $\dot{\phi}$ can never change sign.

The expression (12.4) shows that the radial part of the motion can be regarded as taking place in one dimension in a field where the "effective potential energy" is

$$U_{\text{eff}} = U(r) + M^2/2mr^2. \tag{12.8}$$

The quantity $M^2/2mr^2$ is called the *centrifugal energy*. The values of r for which

$$U(r) + M^2/2mr^2 = E \tag{12.9}$$

determine the limits of the motion as regards distance from the centre. When equation (12.9) is satisfied, the radial velocity $\dot{r}$ is zero. This does not mean that the particle comes to rest as in true one-dimensional motion, since the angular velocity $\dot{\phi}$ is not zero. The value $\dot{r} = 0$ indicates a *turning point* of the path, where $r(t)$ begins to decrease instead of increasing, or *vice versa*.

If the range in which r may vary is limited only by the condition $r \geqslant r_{\text{min}}$, the motion is infinite: the particle comes from, and returns to, infinity.

If the range of r has two limits r_{min} and r_{max}, the motion is finite and the path lies entirely within the annulus bounded by the circles $r = r_{\text{max}}$ and $r = r_{\text{min}}$. This does not mean, however, that the path must be a closed curve. During the time in which r varies from r_{max} to r_{min} and back, the radius vector turns through an angle $\Delta\phi$ which, according to (12.7), is given by

$$\Delta\phi = 2\int_{r_{min}}^{r_{max}} \frac{M\,\mathrm{d}r/r^2}{\sqrt{[2m(E-U)-M^2/r^2]}}. \tag{12.10}$$

The condition for the path to be closed is that this angle should be a rational fraction of 2π, i.e. that $\Delta\phi = 2\pi n_1/n_2$, where n_1 and n_2 are integers. In that case, after n_2 periods, the position vector of the particle will have made n_1 complete revolutions and will occupy its original position, so that the path is closed.

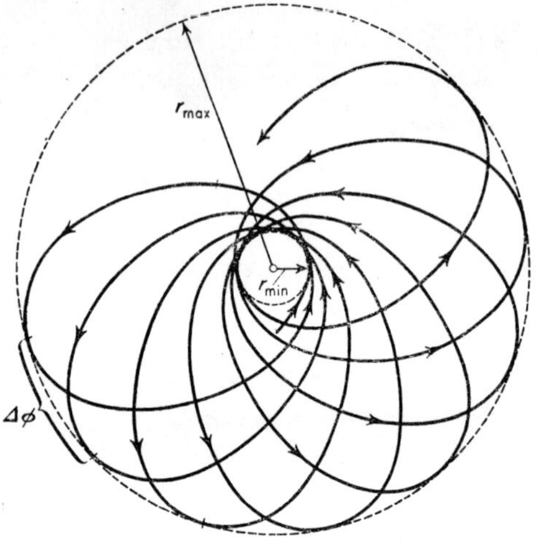

FIG. 6

Such cases are exceptional, however, and when the form of $U(r)$ is arbitrary the angle $\Delta\phi$ is not a rational fraction of 2π. In general, therefore, the path of a particle executing a finite motion is not closed. It passes through the minimum and maximum distances an infinity of times, and after infinite time it covers the entire annulus between the two bounding circles. The path shown in Fig. 6 is an example.

There are only two types of central field in which all finite motions take place in closed paths. They are those in which the potential energy of the particle varies as $1/r$ or as r^2. The former case is discussed in §13; the latter is that of the *space oscillator* (see §19, Problem 3).

§13. Kepler's problem

An important class of central fields is formed by those in which the potential energy is inversely proportional to r, and the force accordingly inversely proportional to r^2. They include the fields of Newtonian gravitational attraction and of Coulomb electrostatic interaction; the latter may be either attractive or repulsive.

Let us first consider an attractive field, where

$$U = -\alpha/r \tag{13.1}$$

with α a positive constant. The "effective" potential energy

$$U_{\text{eff}} = -\frac{\alpha}{r} + \frac{M^2}{2mr^2} \tag{13.2}$$

is of the form shown in Fig. 7. As $r \to 0$, U_{eff} tends to $+\infty$, and as $r \to \infty$ it tends to zero from negative values; for $r = M^2/m\alpha$ it has a minimum value

$$U_{\text{eff, min}} = -m\alpha^2/2M^2. \tag{13.3}$$

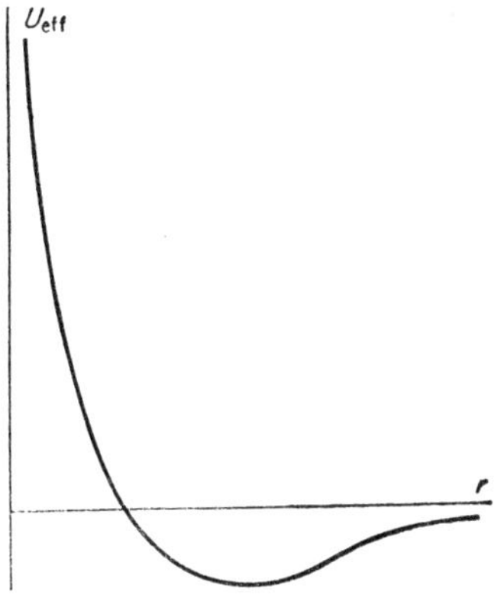

FIG. 7

It is seen at once from Fig. 7 that the motion is finite for $E < 0$ and infinite for $E \geqslant 0$.

The shape of the path is obtained from the general formula (12.7). Substituting there $U = -\alpha/r$ and effecting the elementary integration, we have

$$\phi = \cos^{-1}\frac{(M/r)-(m\alpha/M)}{\sqrt{\left(2mE+\frac{m^2\alpha^2}{M^2}\right)}}+\text{constant}.$$

Taking the origin of ϕ such that the constant is zero, and putting

$$p = M^2/m\alpha, \quad e = \sqrt{[1+(2EM^2/m\alpha^2)]}, \tag{13.4}$$

we can write the equation of the path as

$$p/r = 1+e\cos\phi. \tag{13.5}$$

This is the equation of a conic section with one focus at the origin; $2p$ is called the *latus rectum* of the orbit and e the *eccentricity*. Our choice of the origin of ϕ is seen from (13.5) to be such that the point where $\phi = 0$ is the point nearest to the origin.

In the equivalent problem of two particles interacting according to the law (13.1), the orbit of each particle is a conic section, with one focus at the centre of mass of the two particles.

It is seen from (13.4) that, if $E < 0$, then the eccentricity $e < 1$, i.e. the orbit is an ellipse (Fig. 8) and the motion is finite, in accordance with what has been said earlier in this section. According to the for-

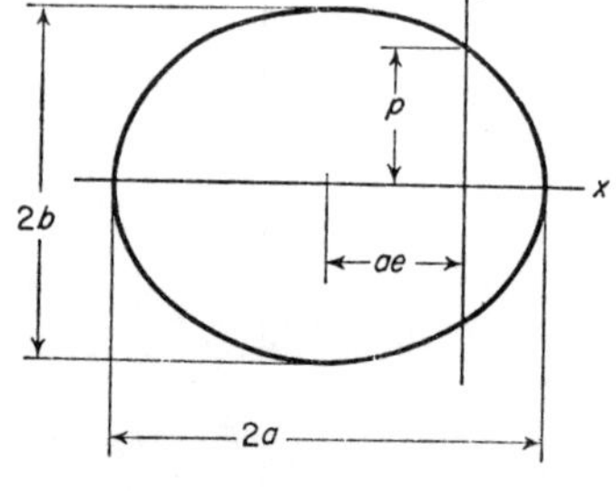

FIG. 8

mulae of analytical geometry, the major and minor semi-axes of the ellipse are

$$a = p/(1-e^2) = \alpha/2|E|, \quad b = p/\sqrt{(1-e^2)} = M/\sqrt{(2m|E|)}. \quad (13.6)$$

The least possible value of the energy is (13.3), and then $e = 0$, i.e. the ellipse becomes a circle. It may be noted that the major axis of the ellipse depends only on the energy of the particle, and not on its angular momentum. The least and greatest distances from the centre of the field (the focus of the ellipse) are

$$r_{min} = p/(1+e) = a(1-e), \quad r_{max} = p/(1-e) = a(1+e). \quad (13.7)$$

These expressions, with a and e given by (13.6) and (13.4), can, of course, also be obtained directly as the roots of the equation $U_{eff}(r) = E$.

The period T of revolution in an elliptical orbit is conveniently found by using the law of conservation of angular momentum in the form of the area integral (12.3). Integrating this equation with respect to time from zero to T, we have $2mf = TM$, where f is the area of the orbit. For an ellipse $f = \pi ab$, and by using the formulae (13.6) we find

$$\begin{aligned} T &= 2\pi a^{3/2}\sqrt{(m/\alpha)} \\ &= \pi\alpha\sqrt{(m/2|E|^3)}. \end{aligned} \quad (13.8)$$

The proportionality between the square of the period and the cube of the linear dimension of the orbit is called *Kepler's third law*. It may also be noted that the period depends only on the energy of the particle.

For $E \geqslant 0$ the motion is infinite. If $E > 0$, the eccentricity $e > 1$, i.e. the path is a hyperbola with the origin as internal focus (Fig. 9). The least distance from the focus is

$$r_{min} = p/(e+1) = a(e-1), \quad (13.9)$$

where $a = p/(e^2-1) = \alpha/2E$ is the "semi-axis" of the hyperbola.

If $E = 0$, the eccentricity $e = 1$, and the particle moves in a parabola with least distance $r_{min} = \frac{1}{2}p$. This case occurs if the particle starts from rest at infinity.

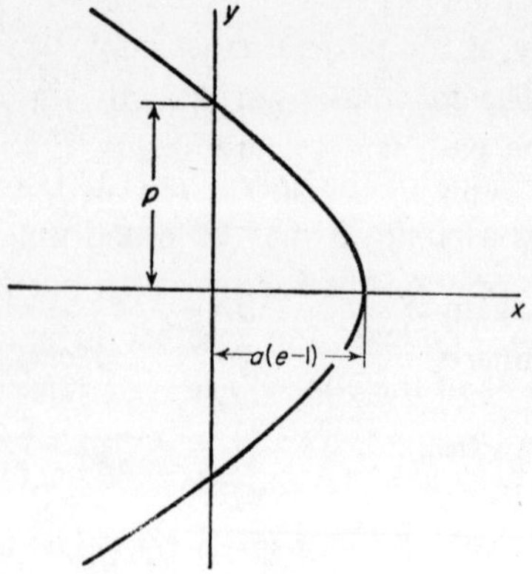

FIG. 9

Let us now consider motion in a repulsive field, where

$$U = \alpha/r \quad (\alpha > 0). \tag{13.10}$$

Here the effective potential energy is

$$U_{\text{eff}} = \frac{\alpha}{r} + \frac{M^2}{2mr^2}$$

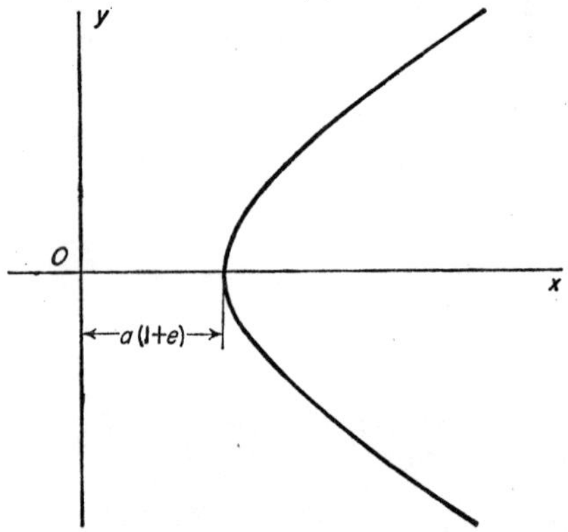

FIG. 10

and decreases monotonically from $+\infty$ to zero as r varies from zero to infinity. The energy of the particle must be positive, and the motion is always infinite. The calculations are exactly similar to those for the attractive field. The path is a hyperbola (or, if $E = 0$, a parabola):

$$p/r = -1+e\cos\phi, \tag{13.11}$$

where p and e are again given by (13.4). The path passes the centre of the field in the manner shown in Fig. 10. The least distance is

$$r_{\min} = p/(e-1) = a(e+1). \tag{13.12}$$

CHAPTER 4

COLLISIONS BETWEEN PARTICLES

§14. Elastic collisions

In many cases the laws of conservation of momentum and energy alone can be used to obtain important results concerning the properties of various mechanical processes. It should be noted that these properties are independent of the particular type of interaction between the particles involved.

Let us consider an *elastic* collision between two particles, i.e. one which involves no change in their internal state. When the law of conservation of energy is applied to such a collision, the internal energy of the particles may be neglected.

The *laboratory system* is a frame of reference in which one of the particles (m_2, say) is at rest before the collision and the other (m_1) moves with velocity $\mathbf{v}$. The collision is most simply described, however, in another frame, where the centre of mass of the two particles is at rest (the *centre-of-mass system*); the values of quantities in this system will be distinguished by the suffix 0. The velocities of the particles before the collision are related to the velocity $\mathbf{v}$ in the laboratory system by $\mathbf{v}_{10} = m_2\mathbf{v}/(m_1+m_2)$, $\mathbf{v}_{20} = -m_1\mathbf{v}/(m_1+m_2)$; see (11.2).

Because of the law of conservation of momentum, the momenta of the two particles remain equal and opposite after the collision, and are also unchanged in magnitude, by the law of conservation of energy. Thus, in the centre-of-mass system the collision simply rotates the velocities, which remain opposite in direction and unchanged in magnitude. If we denote by $\mathbf{n}_0$ a unit vector in the direction of the velocity of the particle m_1 after the collision, then the velocities of the two particles after the collision (distinguished by primes) are

$$\mathbf{v}'_{10} = m_2 v \mathbf{n}_0/(m_1+m_2), \quad \mathbf{v}'_{20} = -m_1 v \mathbf{n}_0/(m_1+m_2). \tag{14.1}$$

In order to return to the laboratory system, we must add to these expressions the velocity $\mathbf{V}$ of the centre of mass. The velocities in the laboratory system after the collision are therefore

$$\left.\begin{aligned} \mathbf{v}'_1 &= m_2 v \mathbf{n}_0/(m_1+m_2)+m_1\mathbf{v}/(m_1+m_2), \\ \mathbf{v}'_2 &= -m_1 v \mathbf{n}_0/(m_1+m_2)+m_1\mathbf{v}/(m_1+m_2). \end{aligned}\right\} \tag{14.2}$$

No further information about the collision can be obtained from the laws of conservation of momentum and energy. The direction of the vector $\mathbf{n}_0$ depends on the law of interaction of the particles and on their relative position during the collision.

The formulae (14.2) for the laboratory system may be interpreted geometrically. Here it is more convenient to use momenta instead of velocities. Multiplying equations (14.2) by m_1 and m_2 respectively, we obtain

$$\left.\begin{aligned} \mathbf{p}'_1 &= mv\mathbf{n}_0+m_1\mathbf{p}_1/(m_1+m_2), \\ \mathbf{p}'_2 &= -mv\mathbf{n}_0+m_2\mathbf{p}_1/(m_1+m_2), \end{aligned}\right\} \tag{14.3}$$

where $m = m_1 m_2/(m_1+m_2)$ is the reduced mass. We draw a circle of radius mv and use the construction shown in Fig. 11. If the unit vector $\mathbf{n}_0$ is along OC, the vectors AC and CB give the momenta $\mathbf{p}'_1$ and $\mathbf{p}'_2$ respectively. When $\mathbf{p}_1$ is given, the radius of the circle and the points A and B are fixed, but the point C may be anywhere on the circle. The point A lies inside or outside the circle, according as $m_1 < m_2$ or $m_1 > m_2$. The corresponding diagrams are shown in Figs. 11a, b. The angles θ_1 and θ_2 in these diagrams are the angles between the directions of motion after the collision and the direction of impact (i.e. of $\mathbf{p}_1$). The angle at the centre, denoted by χ, which gives the

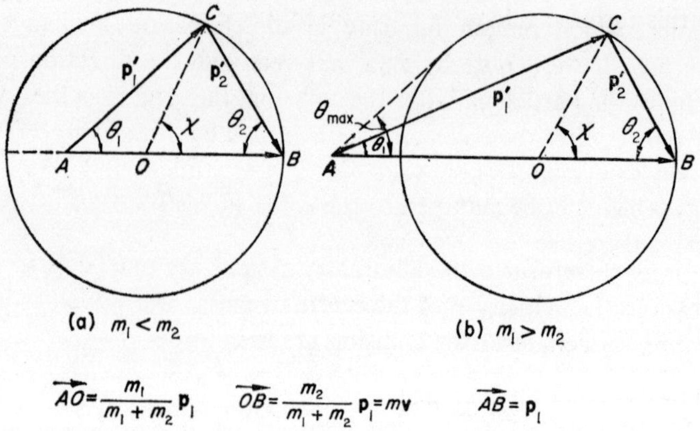

FIG. 11

direction of $\mathbf{n}_0$, is the angle through which the direction of motion of m_1 is turned in the centre-of-mass system. It is evident from the figure that θ_1 and θ_2 can be expressed in terms of χ by

$$\tan\theta_1 = \frac{m_2 \sin\chi}{m_1+m_2\cos\chi}, \qquad \theta_2 = \tfrac{1}{2}(\pi-\chi). \tag{14.4}$$

We may give also the formulae for the magnitudes of the velocities of the two particles after the collision, likewise expressed in terms of χ:

$$v_1' = \frac{\sqrt{(m_1^2+m_2^2+2m_1m_2\cos\chi)}}{m_1+m_2}v, \qquad v_2' = \frac{2m_1 v}{m_1+m_2}\sin\tfrac{1}{2}\chi. \tag{14.5}$$

The sum $\theta_1+\theta_2$ is the angle between the directions of motion of the particles after the collision. Evidently $\theta_1+\theta_2 > \frac{1}{2}\pi$ if $m_1 < m_2$, and $\theta_1+\theta_2 < \frac{1}{2}\pi$ if $m_1 > m_2$.

When the two particles are moving afterwards in the same or in opposite directions (head-on collision), we have $\chi = \pi$, i.e. the point C lies on the diameter through A, and is on OA (Fig. 11b; $\mathbf{p}_1'$ and $\mathbf{p}_2'$ in the same direction) or on OA produced (Fig. 11a; $\mathbf{p}_1'$ and $\mathbf{p}_2'$ in opposite directions).

In this case thc velocities after the collision are

$$\mathbf{v}_1' = \frac{m_1-m_2}{m_1+m_2}\mathbf{v}, \qquad \mathbf{v}_2' = \frac{2m_1}{m_1+m_2}\mathbf{v}. \tag{14.6}$$

This value of $\mathbf{v}_2'$ has the greatest possible magnitude, and the maximum energy which can be acquired in the collision by a particle originally at rest is therefore

$$E'_{2\,\max} = \tfrac{1}{2}\, m_2 v'^2_{2\,\max} = \frac{4m_1m_2}{(m_1+m_2)^2} E_1, \tag{14.7}$$

where $E_1 = \frac{1}{2}m_1v_1^2$ is the initial energy of the incident particle.

If $m_1 < m_2$, the velocity of m_1 after the collision can have any

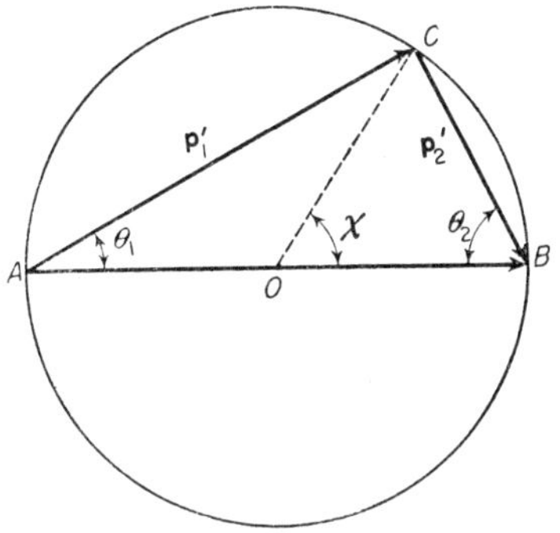

FIG. 12

direction. If $m_1 > m_2$, however, this particle can be deflected only through an angle not exceeding $\theta_{\max}$ from its original direction; this maximum value of θ_1 corresponds to the position of C for which AC is a tangent to the circle (Fig. 11b). Evidently

$$\sin\theta_{\max} = OC/OA = m_2/m_1. \tag{14.8}$$

The collision of two particles of equal mass, of which one is initially at rest, is especially simple. In this case both B and A lie on the circle (Fig. 12).

Then

$$\theta_1 = \tfrac{1}{2}\chi, \qquad \theta_2 = \tfrac{1}{2}(\pi - \chi), \tag{14.9}$$

$$v'_1 = v \cos \tfrac{1}{2}\chi, \quad v'_2 = v \sin \tfrac{1}{2}\chi. \tag{14.10}$$

After the collision the particles move at right angles to each other.

§15. Scattering

As already mentioned in §14, a complete calculation of the result of a collision between two particles (i.e. the determination of the angle χ) requires the solution of the equations of motion for the particular law of interaction involved.

We shall first consider the equivalent problem of the deflection of a single particle of mass m moving in a field $U(r)$ whose centre is at rest

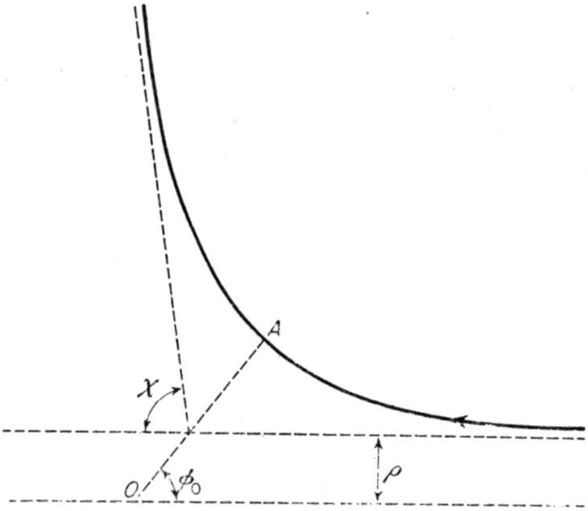

FIG. 13

(and is at the centre of mass of the two particles in the original problem).

The path of a particle in a central field is symmetrical about a line from the centre to the nearest point in the orbit (OA in Fig. 13). Hence the two asymptotes to the orbit make equal angles (ϕ_0, say) with this line. The angle χ through which the particle is deflected as it passes the centre is seen from Fig. 13 to be

$$\chi = |\pi - 2\phi_0|. \tag{15.1}$$

The angle ϕ_0 itself is given, according to (12.7), by

$$\phi_0 = \int_{r_{min}}^{\infty} \frac{(M/r^2)\,\mathrm{d}r}{\sqrt{\{2m[E-U(r)]-M^2/r^2\}}}, \tag{15.2}$$

taken between the nearest approach to the centre and infinity. It should be recalled that r_{min} is a zero of the radicand.

For an infinite motion, such as that considered here, it is convenient to use instead of the constants E and M the velocity v_∞ of the particle at infinity and the *impact parameter* ϱ. The latter is the length of the perpendicular from the centre O to the direction of $\mathbf{v}_\infty$, i.e. the distance at which the particle would pass the centre if there were no field of force (Fig. 13). The energy and the angular momentum are given in terms of these quantities by

$$E = \tfrac{1}{2}mv_\infty^2, \qquad M = m\varrho v_\infty, \tag{15.3}$$

and formula (15.2) becomes

$$\phi_0 = \int_{r_{min}}^{\infty} \frac{(\varrho/r^2)\,\mathrm{d}r}{\sqrt{[1-(\varrho^2/r^2)-(2U/mv_\infty^2)]}}. \tag{15.4}$$

Together with (15.1), this gives χ as a function of ϱ.

In physical applications we are usually concerned not with the deflection of a single particle but with the *scattering* of a beam of identical particles incident with uniform velocity $\mathbf{v}_\infty$ on the scattering centre. The different particles in the beam have different impact parameters and are therefore scattered through different angles χ. Let $\mathrm{d}N$ be the number of particles scattered per unit time through angles between χ and $\chi+\mathrm{d}\chi$. This number itself is not suitable for describing the scattering process, since it is proportional to the density of the incident beam. We therefore use the ratio

$$\mathrm{d}\sigma = \mathrm{d}N/n, \tag{15.5}$$

where n is the number of particles passing in unit time through unit area of the beam cross-section (the beam being assumed uniform over its cross-section). This ratio has the dimensions of area and is called the *(effective) scattering cross-section*. It is entirely determined by the form of the scattering field and is the most important characteristic of the scattering process.

We shall suppose that the relation between χ and ϱ is one-to-one; this is so if the angle of scattering is a monotonically decreasing function of the impact parameter. In that case, only those particles whose impact parameters lie between $\varrho(\chi)$ and $\varrho(\chi)+\mathrm{d}\varrho(\chi)$ are scattered at angles between χ and $\chi+\mathrm{d}\chi$. The number of such particles is equal to the product of n and the area between two circles of radii ϱ and $\varrho+\mathrm{d}\varrho$, i.e. $\mathrm{d}N = 2\pi\varrho\ \mathrm{d}\varrho\cdot n$. The effective cross-section is therefore

$$\mathrm{d}\sigma = 2\pi\varrho\,\mathrm{d}\varrho. \tag{15.6}$$

In order to find the dependence of $\mathrm{d}\sigma$ on the angle of scattering, we need only rewrite (15.6) as

$$\mathrm{d}\sigma = 2\pi\varrho(\chi)|\mathrm{d}\varrho(\chi)/\mathrm{d}\chi|\,\mathrm{d}\chi. \tag{15.7}$$

Here we use the modulus of the derivative $\mathrm{d}\varrho/\mathrm{d}\chi$, since the derivative may be (and usually is) negative. Often $\mathrm{d}\sigma$ is referred to the solid angle element $\mathrm{d}o$ instead of the plane angle element $\mathrm{d}\chi$. The solid angle between cones with vertical angles χ and $\chi+\mathrm{d}\chi$ is $\mathrm{d}o = 2\pi \sin\chi\,\mathrm{d}\chi$. Hence we have from (15.7)

$$\mathrm{d}\sigma = \frac{\varrho(\chi)}{\sin\chi}\left|\frac{\mathrm{d}\varrho}{\mathrm{d}\chi}\right|\mathrm{d}o. \tag{15.8}$$

Returning now to the problem of the scattering of a beam of particles, not by a fixed centre of force, but by other particles initially at rest, we can say that (15.7) gives the effective cross-section as a function of the angle of scattering in the centre-of-mass system. To find the corresponding expression as a function of the scattering angle θ in the laboratory system, we must express χ in (15.7) in terms of θ by means of formulae (14.4). This gives expressions for both the scattering cross-section for the incident beam of particles (χ in terms of θ_1) and that for the particles initially at rest (χ in terms of θ_2).

PROBLEMS

PROBLEM 1. Determine the effective cross-section for scattering of particles from a perfectly rigid sphere of radius a (i.e. when the interaction is such that $U = \infty$ for $r < a$ and $U = 0$ for $r > a$).

SOLUTION. Since a particle moves freely outside the sphere and cannot penetrate into it, the path consists of two straight lines symmetrical about the radius to the point where the particle strikes the sphere (Fig. 14). It is evident from Fig. 14 that

$$\varrho = a \sin \phi_0 = a \sin \tfrac{1}{2}(\pi - \chi) = a \cos \tfrac{1}{2}\chi.$$

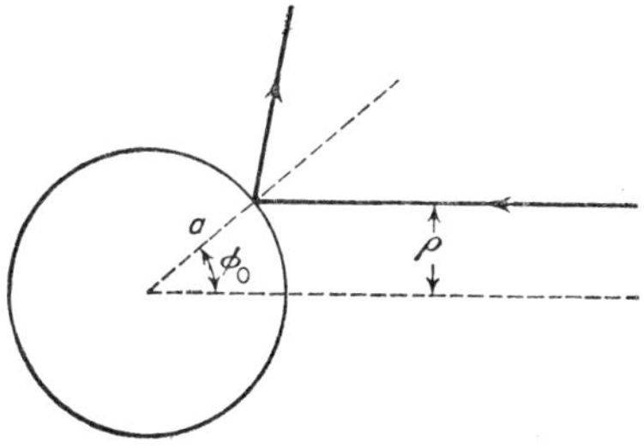

FIG. 14

Substituting in (15.7) or (15.8), we have

$$d\sigma = \tfrac{1}{2}\pi a^2 \sin \chi \, d\chi = \tfrac{1}{4}a^2 \, do, \tag{1}$$

i.e. the scattering is isotropic in the centre-of-mass system. On integrating $d\sigma$ over all angles, we find that the total cross-section $\sigma = \pi a^2$, in accordance with the fact that the "impact area" which the particle must strike in order to be scattered is simply the cross-sectional area of the sphere.

PROBLEM 2. Express the effective cross-section (Problem 1) as a function of the energy ε lost by a scattered particle.

SOLUTION. The energy lost by a particle of mass m_1 is equal to that gained by the sphere of mass m_2. From (14.5) and (14.7), $\varepsilon = E_2' = [2m_1^2 m_2/(m_1+m_2)^2] v_\infty^2 \sin^2 \tfrac{1}{2}\chi = \varepsilon_{max} \sin^2 \tfrac{1}{2}\chi$, whence $d\varepsilon = \tfrac{1}{2}\varepsilon_{max} \sin \chi \, d\chi$; substituting in (1), Problem 1, we have $d\sigma = \pi a^2 \, d\varepsilon/\varepsilon_{max}$. The scattered particles are uniformly distributed with respect to ε in the range from zero to ε_{max}.

PROBLEM 3. Determine the effective cross-section for particles of mass m_1 to strike a sphere of mass m_2 and radius R to which they are attracted in accordance with Newton's law.

SOLUTION. The condition for a particle to reach the sphere is that $r_{min} < R$, where r_{min} is the point on the path which is nearest to the centre of the sphere. The greatest possible value of ϱ is given by $r_{\min} = R$; this is equivalent to $U_{\text{eff}}(R) = E$ or $\frac{1}{2}m_1 v_\infty^2 \varrho_{\max}^2/R^2 - \alpha/R = \frac{1}{2}m_1 v_\infty^2$, where $\alpha = \gamma m_1 m_2$ (γ being the gravitational constant) and we have put $m \approx m_1$ on the assumption that $m_2 \gg m_1$. Solving for $\varrho_{\max}^2$, we finally obtain $\sigma = \pi R^2(1+2\gamma m_2/Rv_\infty^2)$. When $v_\infty \to \infty$ the effective cross-section tends, of course, to the geometrical cross-section of the sphere.

§16. Rutherford's formula

One of the most important applications of the formulae derived above is to the scattering of charged particles in a Coulomb field. Putting in (15.4) $U = \alpha/r$ and effecting the elementary integration, we obtain

$$\phi_0 = \cos^{-1}\frac{\alpha/mv_\infty^2\varrho}{\sqrt{[1+(\alpha/mv_\infty^2\varrho)^2]}},$$

whence $\varrho^2 = (\alpha^2/m^2v_\infty^4)\tan^2\phi_0$, or, putting $\phi_0 = \frac{1}{2}(\pi-\chi)$ from (15.1),

$$\varrho^2 = (\alpha^2/m^2v_\infty^4)\cot^2\tfrac{1}{2}\chi. \tag{16.1}$$

Differentiating this expression with respect to χ and substituting in (15.7) or (15.8) gives

$$\mathrm{d}\sigma = \pi(\alpha/mv_\infty^2)^2\cos\tfrac{1}{2}\chi\,\mathrm{d}\chi/\sin^3\tfrac{1}{2}\chi \tag{16.2}$$

or

$$\mathrm{d}\sigma = (\alpha/2mv_\infty^2)^2\,\mathrm{d}o/\sin^4\tfrac{1}{2}\chi. \tag{16.3}$$

This is *Rutherford's formula*. It may be noted that the effective cross-section is independent of the sign of α, so that the result is equally valid for repulsive and attractive Coulomb fields.

Formula (16.3) gives the effective cross-section in the frame of reference in which the centre of mass of the colliding particles is at rest. The transformation to the laboratory system is effected by means of formulae (14.4). For particles initially at rest we substitute $\chi = \pi - 2\theta_2$ in (16.2) and obtain

$$\begin{aligned}\mathrm{d}\sigma_2 &= 2\pi(\alpha/mv_\infty^2)^2\sin\theta_2\,\mathrm{d}\theta_2/\cos^3\theta_2\\ &= (\alpha/mv_\infty^2)^2\,\mathrm{d}o_2/\cos^3\theta_2.\end{aligned} \tag{16.4}$$

The same transformation for the incident particles leads, in general, to a very complex formula, and we shall merely note two particular cases.

If the mass m_2 of the scattering particle is large compared with the mass m_1 of the scattered particle, then $\chi \approx \theta_1$ and $m \approx m_1$, so that

$$d\sigma_1 = (\alpha/4E_1)^2 \, do_1/\sin^4 \tfrac{1}{2}\theta_1, \tag{16.5}$$

where $E_1 = \frac{1}{2}m_1 v_\infty^2$ is the energy of the incident particle.

If the masses of the two particles are equal ($m_1 = m_2$, $m = \frac{1}{2}m_1$), then by (14.9) $\chi = 2\theta_1$, and substitution in (16.2) gives

$$\begin{aligned} d\sigma_1 &= 2\pi(\alpha/E_1)^2 \cos\theta_1 \, d\theta_1/\sin^3\theta_1 \\ &= (\alpha/E_1)^2 \cos\theta_1 \, do_1/\sin^4\theta_1. \end{aligned} \tag{16.6}$$

If the particles are entirely identical, that which was initially at rest cannot be distinguished after the collision. The total effective cross-section for all particles is obtained by adding $d\sigma_1$ and $d\sigma_2$, and replacing θ_1 and θ_2 by their common value θ:

$$d\sigma = (\alpha/E_1)^2 \left(\frac{1}{\sin^4\theta} + \frac{1}{\cos^4\theta} \right) \cos\theta \, do. \tag{16.7}$$

Let us return to the general formula (16.2) and use it to determine the distribution of the scattered particles with respect to the energy lost in the collision. When the masses of the scattered (m_1) and scattering (m_2) particles are arbitrary, the velocity acquired by the latter is given in terms of the angle of scattering in the centre-of-mass system by $v_2' = [2m_1/(m_1+m_2)]v_\infty \sin \frac{1}{2}\chi$; see (14.5). The energy acquired by m_2 and lost by m_1 is therefore $\varepsilon = \frac{1}{2}m_2 v_2'^2 = (2m^2/m_2)v_\infty^2 \sin^2 \frac{1}{2}\chi$. Expressing $\sin \frac{1}{2}\chi$ in terms of ε and substituting in (16.2), we obtain

$$d\sigma = 2\pi(\alpha^2/m_2 v_\infty^2) \, d\varepsilon/\varepsilon^2. \tag{16.8}$$

This is the required formula: it gives the effective cross-section as a function of the energy loss ε, which takes values from zero to $\varepsilon_{max} = 2m^2 v_\infty^2/m_2$.

CHAPTER 5

SMALL OSCILLATIONS

§17. Free oscillations in one dimension

A very common form of motion of mechanical systems is what are called *small oscillations* of a system about a position of stable equilibrium. We shall consider first of all the simplest case, that of a system with only one degree of freedom.

Stable equilibrium corresponds to a position of the system in which its potential energy $U(q)$ is a minimum. A movement away from this position results in the setting up of a force $-\mathrm{d}U/\mathrm{d}q$ which tends to return the system to equilibrium. Let the equilibrium value of the generalised coordinate q be q_0. For small deviations from the equilibrium position, it is sufficient to retain the first non-vanishing term in the expansion of the difference $U(q)-U(q_0)$ in powers of $q-q_0$. In general this is the second-order term: $U(q)-U(q_0) \approx \frac{1}{2}k(q-q_0)^2$, where k is a positive coefficient, the value of the second derivative $U''(q)$ for $q = q_0$. We shall measure the potential energy from its minimum value, i.e. put $U(q_0) = 0$, and use the symbol

$$x = q-q_0 \tag{17.1}$$

for the deviation of the coordinate from its equilibrium value. Thus

$$U(x) = \tfrac{1}{2}kx^2. \tag{17.2}$$

The kinetic energy of a system with one degree of freedom is in general of the form $\frac{1}{2}a(q)\dot{q}^2 = \frac{1}{2}a(q)\dot{x}^2$. In the same approximation, it is sufficient to replace the function $a(q)$ by its value at $q = q_0$. Putting for brevity† $a(q_0) = m$, we have the following expression for the Lagrangian of a system executing small oscillations in one dimension:‡

$$L = \tfrac{1}{2}m\dot{x}^2 - \tfrac{1}{2}kx^2. \tag{17.3}$$

The corresponding equation of motion is

$$m\ddot{x} + kx = 0, \tag{17.4}$$

or

$$\ddot{x} + \omega^2 x = 0, \tag{17.5}$$

where

$$\omega = \sqrt{(k/m)}. \tag{17.6}$$

Two independent solutions of the linear differential equation (17.5) are $\cos \omega t$ and $\sin \omega t$, and its general solution is therefore

$$x = c_1 \cos \omega t + c_2 \sin \omega t. \tag{17.7}$$

This expression can also be written

$$x = a \cos (\omega t + \alpha). \tag{17.8}$$

Since $\cos(\omega t + \alpha) = \cos \omega t \cos \alpha - \sin \omega t \sin \alpha$, a comparison with (17.7) shows that the arbitrary constants a and α are related to c_1 and c_2 by

$$a = \sqrt{(c_1^2 + c_2^2)}, \quad \tan \alpha = -c_2/c_1. \tag{17.9}$$

Thus, near a position of stable equilibrium, a system excutes harmonic oscillations. The coefficient a of the periodic factor in (17.8) is called the *amplitude* of the oscillations, and the argument of the cosine is their *phase*; α is the initial value of the phase, and evidently depends on the choice of the origin of time. The quantity ω is called the *angular frequency* of the oscillations; in theoretical physics, however, it is usually called simply the *frequency*, and we shall use this name henceforward.

The frequency is a fundamental characteristic of the oscillations, and is independent of the initial conditions of the motion. According to

---

† It should be noticed that m is the mass only if x is the Cartesian coordinate.

‡ Such a system is often called a *one-dimensional oscillator*.

formula (17.6) it is entirely determined by the properties of the mechanical system itself. It should be emphasised, however, that this property of the frequency depends on the assumption that the oscillations are small, and ceases to hold in higher approximations. Mathematically, it depends on the fact that the potential energy is a quadratic function of the coordinate.

The energy of a system executing small oscillations is $E = \frac{1}{2}m\dot{x}^2 + \frac{1}{2}kx^2 = \frac{1}{2}m(\dot{x}^2+\omega^2x^2)$ or, substituting (17.8),

$$E = \tfrac{1}{2}m\omega^2a^2. \tag{17.10}$$

It is proportional to the square of the amplitude.

The time dependence of the coordinate of an oscillating system is often conveniently represented as the real part of a complex expression:

$$x = \mathrm{re}\,[A\exp(-i\omega t)], \tag{17.11}$$

where A is a complex constant; putting

$$A = a\exp(-i\alpha), \tag{17.12}$$

we return to the expression (17.8). The constant A is called the *complex amplitude*; its modulus is the ordinary amplitude, and its argument is the initial phase.

The use of exponential factors is mathematically simpler than that of trigonometrical ones because they are unchanged in form by differentiation. So long as all the operations concerned are linear (addition, multiplication by constants, differentiation, integration), we may omit the sign re throughout and take the real part of the final result.

PROBLEMS

PROBLEM 1. Express the amplitude and initial phase of the oscillations in terms of the initial coordinate x_0 and velocity v_0.

SOLUTION. $a = \sqrt{(x_0^2+v_0^2/\omega^2)}$, $\tan\alpha = -v_0/\omega x_0$.

PROBLEM 2. Find the ratio of frequencies ω and ω' of the oscillations of two diatomic molecules consisting of atoms of different isotopes, the masses of the atoms being m_1, m_2 and m_1', m_2'.

SOLUTION. Since the atoms of the isotopes interact in the same way, we have $k = k'$. The coefficients m in the kinetic energies of the molecules are their reduced masses. According to (17.6) we therefore have

$$\frac{\omega'}{\omega} = \sqrt{\frac{m_1 m_2 (m_1' + m_2')}{m_1' m_2' (m_1 + m_2)}}.$$

PROBLEM 3. Find the frequency of oscillations of a particle of mass m which is free to move along a line and is attached to a spring whose other end is fixed at a point A (Fig. 15) at a distance l from the line. A force F is required to extend the spring to length l.

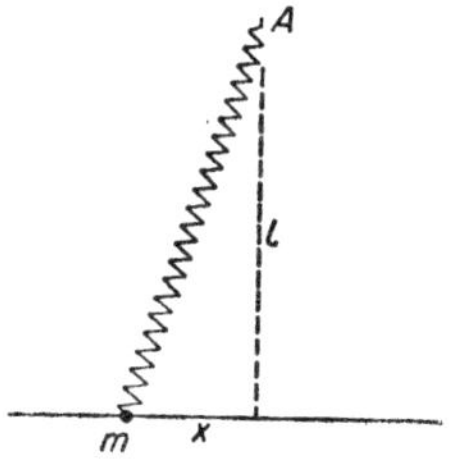

FIG. 15

SOLUTION. The potential energy of the spring is (to within higher-order terms) equal to the force F multiplied by the extension δl of the spring. For $x \ll l$ we have $\delta l = \sqrt{(l^2+x^2)} - l \approx x^2/2l$, so that $U = Fx^2/2l$. Since the kinetic energy is $\frac{1}{2}m\dot{x}^2$, we have $\omega = \sqrt{(F/ml)}$.

§18. Forced oscillations

Let us now consider oscillations of a system on which a variable external force acts. These are called *forced* oscillations, whereas those discussed in §17 are *free* oscillations. Since the oscillations are again supposed small, it is implied that the external field is weak, because otherwise it could cause the displacement x to take too large values.

The system now has, besides the potential energy $\frac{1}{2}kx^2$, the additional potential energy $U_e(x, t)$ resulting from the external field. Expanding this additional term as a series of powers of the small quantity x, we have $U_e(x, t) \approx U_e(0, t) + x[\partial U_e/\partial x]_{x=0}$. The first term is a function of time only, and may therefore be omitted from the Lagrangian, as being the total time derivative of another function of time. In the second term $-[\partial U_e/\partial x]_{x=0}$ is the external "force" acting on the system in the equilibrium position, and is a given function of time, which we denote by $F(t)$. Thus the potential energy involves a further term $-xF(t)$, and the Lagrangian of the system is

$$L = \tfrac{1}{2}m\dot{x}^2 - \tfrac{1}{2}kx^2 + xF(t). \tag{18.1}$$

The corresponding equation of motion is $m\ddot{x}+kx = F(t)$ or

$$\ddot{x}+\omega^2 x = F(t)/m, \tag{18.2}$$

where we have again introduced the frequency ω of the free oscillations.

The general solution of this inhomogeneous linear differential equation with constant coefficients is $x = x_0+x_1$, where x_0 is the general solution of the corresponding homogeneous equation and x_1 is a particular integral of the inhomogeneous equation. In the present case x_0 represents the free oscillations discussed in §17.

Let us consider a case of especial interest, where the external force is itself a simple periodic function of time, of some frequency γ:

$$F(t) = f\cos(\gamma t+\beta). \tag{18.3}$$

We seek a particular integral of equation (18.2) in the form $x_1 = b\cos(\gamma t+\beta)$, with the same periodic factor. Substitution in that equation gives $b = f/m(\omega^2-\gamma^2)$; adding the solution of the homogeneous equation, we obtain the general integral in the form

$$x = a\cos(\omega t+\alpha)+[f/m(\omega^2-\gamma^2)]\cos(\gamma t+\beta). \tag{18.4}$$

The arbitrary constants a and α are found from the initial conditions.

Thus a system under the action of a periodic force executes a motion which is a combination of two oscillations, one with the intrinsic frequency ω of the system and one with the frequency γ of the force.

The solution (18.4) is not valid when *resonance* occurs, i.e. when the frequency γ of the external force is equal to the intrinsic frequency ω of the system. To find the general solution of the equation of motion in this case, we rewrite (18.4) as

$$x = a\cos(\omega t+\alpha)+[f/m(\omega^2-\gamma^2)][\cos(\gamma t+\beta)-\cos(\omega t+\beta)],$$

where a now has a different value. As $\gamma \to \omega$, the second term is indeterminate, of the form 0/0. Resolving the indeterminacy by L'Hospital's rule, we have

$$x = a\cos(\omega t+\alpha)+(f/2m\omega)t\sin(\omega t+\beta). \tag{18.5}$$

Thus the amplitude of oscillations in resonance increases linearly with the time (until the oscillations are no longer small and the whole theory given above becomes invalid).

Let us also ascertain the nature of small oscillations near resonance, when $\gamma = \omega+\varepsilon$ with ε a small quantity. We put the general solution in the complex form

$$\begin{aligned} x &= A\exp(-i\omega t)+B\exp[-i(\omega+\varepsilon)t] \\ &= [A+B\exp(-i\varepsilon t)]\exp(-i\omega t). \end{aligned} \tag{18.6}$$

Since the quantity $A+B\exp(-i\varepsilon t)$ varies only slightly over the period $2\pi/\omega$ of the factor $\exp(-i\omega t)$, the motion near resonance may be regarded as small oscillations of variable amplitude. Denoting this amplitude by C, we have $C = |A+B\exp(-i\varepsilon t)|$. Writing A and B in the form $a\exp(-i\alpha)$ and $b\exp(-i\beta)$ respectively, we obtain

$$C^2 = a^2+b^2+2ab\cos(\varepsilon t+\beta-\alpha). \tag{18.7}$$

Thus the amplitude varies periodically with frequency ε between the limits $|a-b| \leq C \leq a+b$. This phenomenon is called *beats*.

The equation of motion (18.2) can be integrated in a general form for an arbitrary external force $F(t)$. This is easily done by rewriting the equation as

$$\frac{\mathrm{d}}{\mathrm{d}t}(\dot{x}-i\omega x)+i\omega(\dot{x}-i\omega x) = \frac{1}{m}F(t)$$

or

$$\mathrm{d}\xi/\mathrm{d}t+i\omega\xi = F(t)/m, \tag{18.8}$$

where

$$\xi = \dot{x}-i\omega x \tag{18.9}$$

is a complex quantity. Equation (18.8) is of the first order. Its solution when the right-hand side is replaced by zero is $\xi = A\exp(-i\omega t)$ with constant A. As usual, we seek a solution of the inhomogeneous equation in the form $\xi = A(t)\exp(-i\omega t)$, obtaining for the function

$A(t)$ the equation $\dot{A}(t) = F(t)\exp(i\omega t)/m$. Integration gives the solution of (18.8):

$$\xi = \exp(-i\omega t)\left\{\int_0^t \frac{1}{m} F(t)\exp(i\omega t)\,\mathrm{d}t + \xi_0\right\}, \qquad (18.10)$$

where the constant of integration ξ_0 is the value of ξ at the instant $t = 0$. This is the required general solution; the function $x(t)$ is given by the imaginary part of (18.10), divided by $-\omega$.†

The energy of a system executing forced oscillations is naturally not conserved, since the system gains energy from the source of the external field. Let us determine the total energy transmitted to the system during all time, assuming its initial energy to be zero. According to formula (18.10), with the lower limit of integration $-\infty$ instead of zero and with $\xi(-\infty) = 0$, we have for $t \to \infty$

$$|\xi(\infty)|^2 = \frac{1}{m^2}\left|\int_{-\infty}^{\infty} F(t)\exp(i\omega t)\,\mathrm{d}t\right|^2.$$

The energy of the system is

$$E = \tfrac{1}{2}m(\dot{x}^2+\omega^2 x^2) = \tfrac{1}{2}m|\xi|^2. \qquad (18.11)$$

Substituting $|\xi(\infty)|^2$, we obtain the energy transferred:

$$E = \frac{1}{2m}\left|\int_{-\infty}^{\infty} F(t)\exp(i\omega t)\,\mathrm{d}t\right|^2; \qquad (18.12)$$

it is determined by the squared modulus of the Fourier component of the force $F(t)$ whose frequency is the intrinsic frequency of the system.

In particular, if the external force acts only during a time short in comparison with $1/\omega$, we can put $\exp(i\omega t) \approx 1$. Then

$$E = \frac{1}{2m}\left(\int_{-\infty}^{\infty} F(t)\,\mathrm{d}t\right)^2.$$

---

† The force $F(t)$ must, of course, be written in real form.

This result is obvious: it expresses the fact that a force of short duration gives the system a momentum $\int F\,dt$ without bringing about a perceptible displacement.

PROBLEMS

PROBLEM 1. Determine the forced oscillations of a system under a force $F(t)$ of the following forms, if at time $t = 0$ the system is at rest in equilibrium ($x = \dot{x} = 0$): (a) $F = F_0$, a constant, (b) $F = at$, (c) $F = F_0 \exp(-\alpha t)$, (d) $F = F_0 \exp(-\alpha t) \times \cos \beta t$.

SOLUTION. (a) $x = (F_0/m\omega^2)(1 - \cos \omega t)$. The action of the constant force results in a displacement of the position of equilibrium about which the oscillations take place.

(b) $x = (a/m\omega^3)(\omega t - \sin \omega t)$.

(c) $x = [F_0/m(\omega^2+\alpha^2)] [\exp(-\alpha t) - \cos \omega t + (\alpha/\omega) \sin \omega t]$.

(d) $x = F_0\{-(\omega^2+\alpha^2-\beta^2) \cos \omega t + (\alpha/\omega)(\omega^2+\alpha^2+\beta^2) \sin \omega t + {}$
$+ \exp(-\alpha t) [(\omega^2+\alpha^2-\beta^2) \cos \beta t - 2\alpha\beta \sin \beta t]\}/m[(\omega^2+\alpha^2-\beta^2)^2 + 4\alpha^2\beta^2]$.

This last case is conveniently treated by writing the force in the complex form

$$F = F_0 \exp[(-\alpha - i\beta)t].$$

PROBLEM 2. Determine the final amplitude for the oscillations of a system under a force which is zero for $t < 0$, $F_0 t/T$ for $0 < t < T$, and F_0 for $t > T$ (Fig. 16), if up to time $t = 0$ the system is at rest in equilibrium.

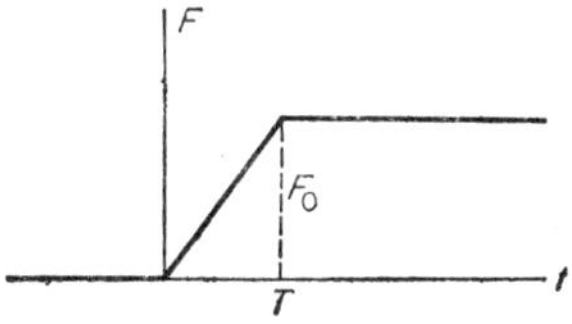

FIG. 16

SOLUTION. During the interval $0 < t < T$ the oscillations are determined by the initial condition as $x = (F_0/mT\omega^3)(\omega t - \sin \omega t)$. For $t > T$ we seek a solution in the form

$$x = c_1 \cos \omega(t-T) + c_2 \sin \omega(t-T) + F_0/m\omega^2.$$

The continuity of x and $\dot{x}$ at $t = T$ gives $c_1 = -(F_0/mT\omega^3) \sin \omega T$, $c_2 = (F_0/mT\omega^3)(1 - \cos \omega T)$. The amplitude is $a = \sqrt{(c_1^2 + c_2^2)} = (2F_0/mT\omega^3) \sin \frac{1}{2}\omega T$. This is the smaller, the more slowly the force F_0 is applied (i.e. the greater T).

PROBLEM 3. The same as Problem 2, but for a constant force F_0 which acts for a finite time T (Fig. 17).

SOLUTION. As in Problem 2, or more simply by using formula (18.10). For $t > T$ we have free oscillations about $x = 0$, and

$$\xi = \frac{F_0}{m}\exp(-i\omega t)\int_0^T \exp(i\omega t)\,dt$$

$$= \frac{iF_0}{\omega m}[1-\exp(i\omega T)]\exp(-i\omega t).$$

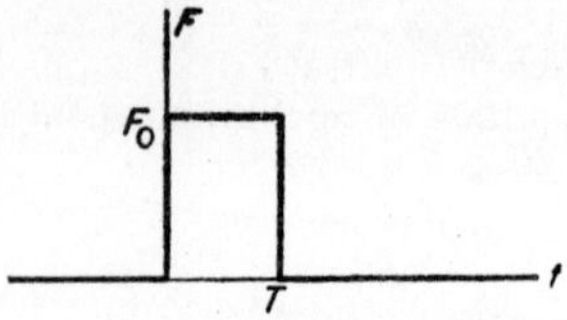

FIG. 17

The squared modulus of ξ gives the amplitude from the relation $|\xi|^2 = a^2\omega^2$. The result is

$$a = (2F_0/m\omega^2)\sin\tfrac{1}{2}\omega T.$$

§19. Oscillations of systems with more than one degree of freedom

The theory of free oscillations of systems with s degrees of freedom is analogous to that given in §17 for the case $s = 1$.

Let the potential energy of the system U as a function of the generalised coordinates q_i ($i = 1, 2, \ldots, s$) have a minimum for $q_i = q_{i0}$. Putting

$$x_i = q_i - q_{i0} \tag{19.1}$$

for the small displacements from equilibrium and expanding U as a function of the x_i as far as the quadratic terms, we obtain the potential energy as a positive definite quadratic form

$$U = \tfrac{1}{2}\sum_{i,k} k_{ik}x_i x_k, \tag{19.2}$$

where we again take the minimum value of the potential energy as zero. Since the coefficients k_{ik} and k_{ki} in (19.2) multiply the same quantity $x_i x_k$, it is clear that they may always be considered equal: $k_{ik} = k_{ki}$.

In the kinetio energy, which has the general form $\frac{1}{2}\Sigma a_{ik}(q)\dot{q}_i\dot{q}_k$ (see (5.5)), we put $q_i = q_{i0}$ in the coefficients a_{ik} and, denoting $a_{ik}(q_0)$ by m_{ik}, obtain the kinetic energy as a positive definite quadratic form

$$\frac{1}{2}\sum_{i,k} m_{ik}\dot{x}_i\dot{x}_k. \tag{19.3}$$

The coefficients m_{ik} also may always be regarded as symmetrical: $m_{ik} = m_{ki}$. Thus the Lagrangian of a system executing small free oscillations is

$$L = \frac{1}{2}\sum_{i,k}(m_{ik}\dot{x}_i\dot{x}_k - k_{ik}x_i x_k). \tag{19.4}$$

Let us now derive the equations of motion. To determine the derivatives involved, we write the total differential of the Lagrangian:

$$\mathrm{d}L = \frac{1}{2}\sum_{i,k}(m_{ik}\dot{x}_i\,\mathrm{d}\dot{x}_k + m_{ik}\dot{x}_k\,\mathrm{d}\dot{x}_i - k_{ik}x_i\,\mathrm{d}x_k - k_{ik}x_k\,\mathrm{d}x_i).$$

Since the value of the sum is obviously independent of the naming of the suffixes, we can interchange i and k in the first and third terms in the parentheses. Using the symmetry of m_{ik} and k_{ik}, we have

$$\mathrm{d}L = \sum(m_{ik}\dot{x}_k\,\mathrm{d}\dot{x}_i - k_{ik}x_k\,\mathrm{d}x_i).$$

Hence

$$\partial L/\partial\dot{x}_i = \sum_k m_{ik}\dot{x}_k, \qquad \partial L/\partial x_i = -\sum_k k_{ik}x_k.$$

Lagrange's equations are therefore

$$\sum_k m_{ik}\ddot{x}_k + \sum_k k_{ik}x_k = 0 \qquad (i = 1, 2, \ldots, s); \tag{19.5}$$

they form a set of s linear homogeneous differential equations with constant coefficients.

As usual, we seek the s unknown functions $x_k(t)$ in the form

$$x_k = A_k \exp(-i\omega t), \tag{19.6}$$

where A_k are some constants to be determined. Substituting (19.6) in the equations (19.5) and cancelling $\exp(-i\omega t)$, we obtain a set of linear homogeneous algebraic equations to be satisfied by the A_k:

$$\sum_k (-\omega^2 m_{ik} + k_{ik}) A_k = 0. \tag{19.7}$$

If this system has non-zero solutions, the determinant of the coefficients must vanish:

$$|k_{ik} - \omega^2 m_{ik}| = 0. \tag{19.8}$$

This is the *characteristic equation* and is of degree s in ω^2. In general, it has s different real positive roots ω_α^2 ($\alpha = 1, 2, \ldots, s$). The quantities ω_α thus determined are the *characteristic frequencies* or *eigenfrequencies* of the system. In particular cases, some of the roots may coincide; the corresponding multiple eigenfrequencies are said to be *degenerate*.

It is evident from physical arguments that the roots of equation (19.8) are real and positive. For the existence of an imaginary part of ω would mean the presence, in the time dependence of the coordinates x_k (19.6), and so of the velocities $\dot{x}_k$, of an exponentially decreasing or increasing factor. Such a factor is inadmissible, since it would lead to a time variation of the total energy $E = U + T$ of the system, which would therefore not be conserved.

The frequencies ω_α having been found, we substitute each of them in equations (19.7) and find the corresponding coefficients A_k. Since these equations are homogeneous, there is an arbitrary common factor in the resulting values. To stress this, we shall write the coefficients A_k for any given frequency ω_α as $\Delta_{k\alpha} C_\alpha$, with a specified set of real constants $\Delta_{k\alpha}$ and an arbitrary (complex) constant C_α which does not depend on k.

The particular solution of the differential equations (19.5) is therefore

$$x_k = \Delta_{k\alpha} C_\alpha \exp(-i\omega_\alpha t).$$

The general solution is the sum of all the particular solutions. Taking the real part, we write

$$x_k = \sum_\alpha \Delta_{k\alpha} Q_\alpha, \tag{19.9}$$

where

$$Q_\alpha = \text{re}\,[C_\alpha \exp(-i\omega_\alpha t)]. \tag{19.10}$$

Thus the time variation of each coordinate of the system is a superposition of s simple periodic oscillations $Q_1, Q_2, \ldots, Q_s$ with arbitrary amplitudes and phases but definite frequencies.

The question naturally arises whether the generalised coordinates can be chosen in such a way that each of them executes only one simple oscillation. The form of the general integral (19.9) points to the answer. For, regarding the s equations (19.9) as a set of equations for s unknowns Q_α, we can express $Q_1, Q_2, \ldots, Q_s$ in terms of the coordinates $x_1, x_2, \ldots, x_s$. The quantities Q_α may therefore be regarded as new generalised coordinates, called *normal coordinates*, and they execute simple periodic oscillations, called *normal oscillations* of the system.

The normal coordinates Q_α are seen from their definition to satisfy the equations

$$\ddot{Q}_\alpha + \omega_\alpha^2 Q_\alpha = 0. \tag{19.11}$$

This means that in normal coordinates the equations of motion become s independent equations. The acceleration in each normal coordinate depends only on the value of that coordinate, and its time dependence is entirely determined by the initial values of the coordinate and of the corresponding velocity. In other words, the normal oscillations of the system are completely independent.

It is evident that the Lagrangian expressed in terms of normal coordinates is a sum of expressions each of which corresponds to oscillation in one dimension with one of the frequencies ω_α, i.e. it is of the form

$$L = \sum_\alpha \tfrac{1}{2} m_\alpha (\dot{Q}_\alpha^2 - \omega_\alpha^2 Q_\alpha^2),$$

where each m_α is a positive constant which can take any value if the $\Delta_{k\alpha}$ in (19.9) are changed by an appropriate factor. The normal coordinates are usually so chosen as to make $m_\alpha = 1$. Then the

Lagrangian of the system is†

$$L = \tfrac{1}{2}\sum_{\alpha}(\dot{Q}_{\alpha}^{2} - \omega_{\alpha}^{2}Q_{\alpha}^{2}). \tag{19.12}$$

If we have a system of interacting particles not in an external field, not all of its degrees of freedom relate to oscillations. A typical example is that of molecules. Besides motions in which the atoms oscillate about their positions of equilibrium in the molecule, the whole molecule can execute translational and rotational motions.

Three degrees of freedom correspond to translational motion, and in general the same number to rotation, so that, of the $3n$ degrees of freedom of a molecule containing n atoms, $3n-6$ correspond to vibration. An exception is formed by molecules in which the atoms are collinear, for which there are only two rotational degrees of freedom (since rotation about the line of atoms is of no significance), and therefore $3n-5$ vibrational degrees of freedom.

The normal vibrations of the molecule may be classified according to the corresponding motion of the atoms on the basis of a consideration of the symmetry of the equilibrium positions of the atoms in the molecule. There is a general method of doing so, based on the use of group theory. Here we shall consider only some elementary examples.

If all n atoms in a molecule lie in one plane, we can distinguish normal vibrations in which the atoms remain in that plane from those where they do not. The number of each kind is readily determined. Since, for motion in a plane, there are $2n$ degrees of freedom, of which two are translational and one rotational, the number of normal vibrations which leave the atoms in the plane is $2n-3$. The remaining $(3n-6)-(2n-3) = n-3$ vibrational degrees of freedom correspond to vibrations in which the atoms move out of the plane.

† When there are degenerate frequencies, the choice of the normal coordinates is not fully determined by this condition. Since those coordinates which have the same ω_α appear in the kinetic and potential energies as sums $\sum Q_{\alpha}^{2}$ and $\sum \dot{Q}_{\alpha}^{2}$ which are transformed in the same way, they can be linearly transformed in any manner which does not alter these sums of squares.

For a linear molecule we can distinguish longitudinal vibrations, which maintain the linear form, from vibrations which bring the atoms out of line. Since a motion of n particles in a line corresponds to n degrees of freedom, of which one is translational, the number of vibrations which leave the atoms in line is $n-1$. Since the total number of vibrational degrees of freedom of a linear molecule is $3n-5$, there are $2n-4$ which bring the atoms out of line. These $2n-4$ vibrations, however, correspond to only $n-2$ different frequencies, since each such vibration can occur in two mutually perpendicular planes through the axis of the molecule. It is evident from symmetry that each such pair of normal vibrations have equal frequencies.

PROBLEMS

PROBLEM 1. Determine the oscillations of a system with two degrees of freedom whose Lagrangian is $L = \frac{1}{2}(\dot{x}^2+\dot{y}^2)-\frac{1}{2}\omega_0^2(x^2+y^2)+\alpha xy$ (two identical one-dimensional systems of eigenfrequency ω_0 coupled by an interaction $-\alpha xy$).

SOLUTION. The equations of motion are $\ddot{x}+\omega_0^2 x = \alpha y$, $\ddot{y}+\omega_0^2 y = \alpha x$. The substitution (19.6) gives

$$A_x(\omega_0^2-\omega^2) = \alpha A_y, \qquad A_y(\omega_0^2-\omega^2) = \alpha A_x. \tag{1}$$

The characteristic equation is $(\omega_0^2-\omega^2)^2 = \alpha^2$, whence $\omega_1^2 = \omega_0^2-\alpha$, $\omega_2^2 = \omega_0^2+\alpha$. For $\omega = \omega_1$, the equations (1) give $A_x = A_y$, and for $\omega = \omega_2$, $A_x = -A_y$. Hence $x = (Q_1+Q_2)/\sqrt{2}$, $y = (Q_1-Q_2)/\sqrt{2}$, the coefficients $1/\sqrt{2}$ resulting from the normalisation of the normal coordinates.

For $\alpha \ll \omega_0^2$ (weak coupling) we have $\omega_1 \approx \omega_0-\frac{1}{2}\alpha/\omega_0$, $\omega_2 \approx \omega_0+\frac{1}{2}\alpha/\omega_0$. The variation of x and y is in this case a superposition of two oscillations with almost equal frequencies, i.e. beats of frequency $\omega_2-\omega_1 = \alpha/\omega_0$ (see §18). The amplitude of y is a minimum when that of x is a maximum, and *vice versa*.

PROBLEM 2. Determine the small oscillations of a coplanar double pendulum (Fig. 1, §5).

SOLUTION. For small oscillations ($\phi_1 \ll 1$, $\phi_2 \ll 1$), the Lagrangian derived in §5, Problem 1, becomes

$$L = \tfrac{1}{2}(m_1+m_2)l_1^2\dot{\phi}_1^2+\tfrac{1}{2}m_2 l_2^2\dot{\phi}_2^2+m_2 l_1 l_2\dot{\phi}_1\dot{\phi}_2-\tfrac{1}{2}(m_1+m_2)gl_1\phi_1^2-\tfrac{1}{2}m_2 g l_2\phi_2^2.$$

The equations of motion are

$$(m_1+m_2)l_1\ddot{\phi}_1+m_2 l_2\ddot{\phi}_2+(m_1+m_2)g\phi_1 = 0, \qquad l_1\ddot{\phi}_1+l_2\ddot{\phi}_2+g\phi_2 = 0.$$

Substitution of (19.6) gives

$$A_1(m_1+m_2)(g-l_1\omega^2)-A_2\omega^2 m_2 l_2 = 0, \qquad -A_1 l_1\omega^2+A_2(g-l_2\omega^2) = 0.$$

The roots of the characteristic equation are

$$\omega_{1,2}^2 = \frac{g}{2m_1 l_1 l_2}\{(m_1+m_2)(l_1+l_2) \pm \sqrt{(m_1+m_2)}\sqrt{[(m_1+m_2)(l_1+l_2)^2 - 4m_1 l_1 l_2]}\}.$$

As $m_1 \to \infty$ the frequencies tend to the values $\sqrt{(g/l_1)}$ and $\sqrt{(g/l_2)}$, corresponding to independent oscillations of the two pendulums.

PROBLEM 3. Find the path of a particle in a central field $U = \frac{1}{2}kr^2$ (called a *space oscillator*).

SOLUTION. As in any central field, the path lies in a plane, which we take as the xy plane. The variation of each coordinate x, y is a simple oscillation with the same frequency $\omega = \sqrt{(k/m)}$: $x = a\cos(\omega t+\alpha)$, $y = b\cos(\omega t+\beta)$, or $x = a\cos\phi$, $y = b\cos(\phi+\delta) = b\cos\delta\cos\phi - b\sin\delta\sin\phi$, where $\phi = \omega t+\alpha$, $\delta = \beta-\alpha$. Solving for $\cos\phi$ and $\sin\phi$ and equating the sum of their squares to unity, we find the equation of the path:

$$\frac{x^2}{a^2} + \frac{y^2}{b^2} - \frac{2xy}{ab}\cos\delta = \sin^2\delta.$$

This is an ellipse with its centre at the origin. When $\delta = 0$ or π, the path degenerates to a segment of a straight line.

§20. Damped oscillations

So far we have implied that all motion takes place in a vacuum, or else that the effect of the surrounding medium on the motion may be neglected. In reality, when a body moves in a medium, the latter exerts a resistance which tends to retard the motion. The energy of the moving body is finally dissipated by being converted into heat.

Motion under these conditions is no longer a purely mechanical process, and allowance must be made for the motion of the medium itself and for the internal thermal state of both the medium and the body. In particular, we cannot in general assert that the acceleration of a moving body is a function only of its coordinates and velocity at the instant considered; that is, there are no equations of motion derivable from the Lagrangian by the methods of mechanics. Thus the problem of the motion of a body in a medium is not one of mechanics alone.

There exists, however, a class of cases where motion in a medium can be approximately described by including certain additional terms in the mechanical equations of motion. Such cases include oscillations with frequencies small compared with those of the dissipative processes in the medium. When this condition is fulfilled we may regard the body as being acted on by a force of *friction* which depends (for a given homogeneous medium) only on its velocity.

If, in addition, this velocity is sufficiently small, then the frictional force can be expanded in powers of the velocity. The zero-order term in the expansion is zero, since no friction acts on a body at rest, and so the first non-vanishing term is proportional to the velocity. Thus the generalised frictional force f_{fr} acting on a system executing small oscillations in one dimension (coordinate x) may be written $f_{fr} = -\alpha\dot{x}$, where α is a positive coefficient and the minus sign indicates that the force acts in the direction opposite to that of the velocity. Adding this force on the right-hand side of the equation of motion, we obtain

$$m\ddot{x} = -kx-\alpha\dot{x}. \tag{20.1}$$

We divide this by m and put

$$k/m = \omega_0^2, \quad \alpha/m = 2\lambda; \tag{20.2}$$

ω_0 is the frequency of free oscillations of the system in the absence of friction, and λ is called the *damping coefficient*.†

Thus the equation is

$$\ddot{x}+2\lambda\dot{x}+\omega_0^2 x = 0. \tag{20.3}$$

We seek a solution $x = \exp(rt)$ and obtain for r the characteristic equation $r^2+2\lambda r+\omega_0^2 = 0$. The general solution of equation (20.3) is

$$x = c_1 \exp(r_1 t)+c_2 \exp(r_2 t), \quad r_{1,2} = -\lambda \pm \sqrt{(\lambda^2-\omega_0^2)}.$$

Two cases must be distinguished. If $\lambda < \omega_0$, we have two complex conjugate values of r. The general solution of the equation of motion can then be written as

$$x = \text{re}\{A \exp[-\lambda t - i\sqrt{(\omega_0^2-\lambda^2)}t]\},$$

† The dimensionless product λT (where $T = 2\pi/\omega$ is the period) is called the *logarithmic damping decrement*.

where A is an arbitrary complex constant, or as

$$x = a\exp(-\lambda t)\cos(\omega t+\alpha), \tag{20.4}$$

with $\omega = \sqrt{(\omega_0^2-\lambda^2)}$ and a and α real constants. The motion described by these formulae consists of *damped oscillations*. It may be regarded as being harmonic oscillations of exponentially decreasing amplitude. The rate of decrease of the amplitude is given by the exponent λ, and the "frequency" ω is less than that of free oscillations in the absence of friction. For $\lambda \ll \omega_0$, the difference between ω and ω_0 is of the second order of smallness. The decrease in frequency as a result of friction is to be expected, since friction retards motion.

If $\lambda \ll \omega_0$, the amplitude of the damped oscillation is almost unchanged during the period $2\pi/\omega$. It is then meaningful to consider the mean values (over the period) of the squared coordinates and velocities, neglecting the change in $\exp(-\lambda t)$ when taking the mean. These mean squares are evidently proportional to $\exp(-2\lambda t)$. Hence the mean energy of the system decreases as

$$\bar{E} = E_0\exp(-2\lambda t), \tag{20.5}$$

where E_0 is the initial value of the energy.

Next, let $\lambda > \omega_0$. Then the values of r are both real and negative. The general form of the solution is

$$x = c_1\exp\{-[\lambda-\sqrt{(\lambda^2-\omega_0^2)}]t\}+c_2\exp\{-[\lambda+\sqrt{(\lambda^2-\omega_0^2)}]t\}. \tag{20.6}$$

We see that in this case, which occurs when the friction is sufficiently strong, the motion consists of a decrease in $|x|$, i.e. an asymptotic approach (as $t \to \infty$) to the equilibrium position. This type of motion is called *aperiodic damping*.

Finally, in the special case where $\lambda = \omega_0$, the characteristic equation has the double root $r = -\lambda$. The general solution of the differential equation is then

$$x = (c_1+c_2t)\exp(-\lambda t). \tag{20.7}$$

This is a special case of aperiodic damping.

For a system with more than one degree of freedom, the generalised frictional forces corresponding to the coordinates x_i are linear functions of the velocities, of the form

$$f_{\mathrm{fr},i} = -\sum_k \alpha_{ik}\dot{x}_k. \qquad (20.8)$$

From purely mechanical arguments we can draw no conclusions concerning the symmetry properties of the coefficients α_{ik} as regards the suffixes i and k, but the methods of statistical physics make it possible to demonstrate that in all cases

$$\alpha_{ik} = \alpha_{ki}. \qquad (20.9)$$

Hence the expressions (20.8) can be written as the derivatives

$$f_{\mathrm{fr},i} = -\partial F/\partial \dot{x}_i \qquad (20.10)$$

of the quadratic form

$$F = \tfrac{1}{2}\sum_{i,k} \alpha_{ik}\dot{x}_i\dot{x}_k, \qquad (20.11)$$

which is called the *dissipative function.*

The forces (20.10) must be added to the right-hand side of Lagrange's equations:

$$\frac{\mathrm{d}}{\mathrm{d}t}\left(\frac{\partial L}{\partial \dot{x}_i}\right) = \frac{\partial L}{\partial x_i} - \frac{\partial F}{\partial \dot{x}_i}. \qquad (20.12)$$

The dissipative function itself has an important physical significance: it gives the rate of dissipation of energy in the system. This is easily seen by calculating the time derivative of the mechanical energy of the system. We have

$$\begin{aligned}\frac{\mathrm{d}E}{\mathrm{d}t} &= \frac{\mathrm{d}}{\mathrm{d}t}\left(\sum_i \dot{x}_i \frac{\partial L}{\partial \dot{x}_i} - L\right)\\ &= \sum_i \dot{x}_i\left(\frac{\mathrm{d}}{\mathrm{d}t}\left[\frac{\partial L}{\partial \dot{x}_i}\right] - \frac{\partial L}{\partial x_i}\right)\\ &= -\sum_i \dot{x}_i \frac{\partial F}{\partial \dot{x}_i}.\end{aligned}$$

Since F is a quadratic function of the velocities, Euler's theorem on homogeneous functions shows that the sum on the right-hand side is equal to $2F$. Thus

$$dE/dt = -2F, \tag{20.13}$$

i.e. the rate of change of the energy of the system is twice the dissipative function. Since dissipative processes lead to loss of energy, it follows that $F > 0$, i.e. the quadratic form (20.11) is positive definite.

§21. Forced oscillations under friction

The theory of forced oscillations under friction is entirely analogous to that given in §18 for oscillations without friction. Here we shall consider in detail the case of a periodic external force, which is of considerable interest.

Adding to the right-hand side of equation (20.1) an external force $f \cos \gamma t$ and dividing by m, we obtain the equation of motion:

$$\ddot{x} + 2\lambda\dot{x} + \omega_0^2 x = (f/m)\cos \gamma t. \tag{21.1}$$

The solution of this equation is more conveniently found in complex form, and so we replace $\cos \gamma t$ on the right by $\exp(-i\gamma t)$:

$$\ddot{x} + 2\lambda\dot{x} + \omega_0^2 x = (f/m)\exp(-i\gamma t).$$

We seek a particular integral in the form $x = B\exp(-i\gamma t)$, obtaining for B the value

$$B = f/m(\omega_0^2 - \gamma^2 - 2i\lambda\gamma). \tag{21.2}$$

Writing $B = b\exp(-i\delta)$, we have

$$b = f/m\sqrt{[(\omega_0^2 - \gamma^2)^2 + 4\lambda^2\gamma^2]}, \qquad \tan\delta = 2\lambda\gamma/(\gamma^2 - \omega_0^2). \tag{21.3}$$

Finally, taking the real part of the expression $B\exp(-i\gamma t) = b\exp[-i(\gamma t + \delta)]$, we find the particular integral of equation (21.1); adding to this the general solution of that equation with zero on the right-hand side (and taking for definiteness the case $\omega_0 > \lambda$), we have

$$x = a\exp(-\lambda t)\cos(\omega t + \alpha) + b\cos(\gamma t + \delta). \tag{21.4}$$

The first term decreases exponentially with time, so that, after a sufficient time, only the second term remains:

$$x = b\cos(\gamma t+\delta). \tag{21.5}$$

The expression (21.3) for the amplitude b of the forced oscillation increases as γ approaches ω_0, but does not become infinite as it does in resonance without friction. For a given amplitude f of the force, the amplitude of the oscillations is greatest when $\gamma = \sqrt{(\omega_0^2-2\lambda^2)}$; for $\lambda \ll \omega_0$, this differs from ω_0 only by a quantity of the second order of smallness.

Let us consider the range near resonance, putting $\gamma = \omega_0+\varepsilon$ with ε small, and suppose also that $\lambda \ll \omega_0$. Then we can approximately put, in (21.2), $\gamma^2-\omega_0^2 = (\gamma+\omega_0)(\gamma-\omega_0) \approx 2\omega_0\varepsilon$, $2i\lambda\gamma \approx 2i\lambda\omega_0$, so that

$$B = -f/2m(\varepsilon+i\lambda)\omega_0 \tag{21.6}$$

or

$$b = f/2m\omega_0\sqrt{(\varepsilon^2+\lambda^2)}, \qquad \tan\delta = \lambda/\varepsilon. \tag{21.7}$$

A property of the phase difference δ between the oscillation and the external force is that it is always negative, i.e. the oscillation "lags behind" the force. Far from resonance on the side $\gamma < \omega_0$, $\delta \to 0$; on the side $\gamma > \omega_0$, $\delta \to -\pi$. The change of δ from zero to $-\pi$ takes place in a frequency range near ω_0 which is narrow (of the order of λ in width); δ passes through $-\frac{1}{2}\pi$ when $\gamma = \omega_0$. In the absence of friction, the phase of the forced oscillation changes discontinuously by π at $\gamma = \omega_0$ (the second term in (18.4) changes sign); when friction is allowed for, this discontinuity is smoothed out.

In steady motion, when the system executes the forced oscillations given by (21.5), its energy remains unchanged. Energy is continually absorbed by the system from the source of the external force and dissipated by friction. Let $I(\gamma)$ be the mean amount of energy absorbed per unit time, which depends on the frequency of the external force. By (20.13) we have $I(\gamma) = 2\bar{F}$, where $\bar{F}$ is the average value (over the period of oscillation) of the dissipative function. For motion in one dimension, the expression (20.11) for the dissipative function becomes $F = \frac{1}{2}\alpha\dot{x}^2 = \lambda m\dot{x}^2$. Substituting (21.5), we have

$$F = \lambda m b^2\gamma^2\sin^2(\gamma t+\delta).$$

The time average of the squared sine is $\frac{1}{2}$, so that

$$I(\gamma) = \lambda m b^2 \gamma^2. \tag{21.8}$$

Near resonance we have, on substituting the amplitude of the oscillation from (21.7),

$$I(\varepsilon) = f^2\lambda/4m(\varepsilon^2+\lambda^2). \tag{21.9}$$

This is called a *dispersion-type* frequency dependence of the absorption. The *half-width* of the resonance curve (Fig. 18) is the value of $|\varepsilon|$ for which $I(\varepsilon)$ is half its maximum value ($\varepsilon = 0$). It is evident from (21.9)

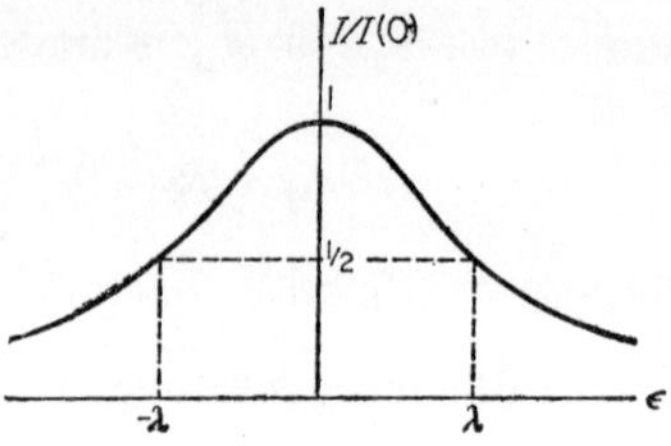

FIG. 18

that in the present case the half-width is just the damping coefficient λ. The height of the maximum is $I(0) = f^2/4m\lambda$, and is inversely proportional to λ. Thus, when the damping coefficient decreases, the resonance curve becomes more peaked. The area under the curve, however, remains unchanged. This area is given by the integral

$$\int_0^\infty I(\gamma)\,\mathrm{d}\gamma = \int_{-\omega_0}^\infty I(\varepsilon)\,\mathrm{d}\varepsilon.$$

Since $I(\varepsilon)$ diminishes rapidly with increasing $|\varepsilon|$, the region where $|\varepsilon|$ is large is of no importance, and the lower limit may be replaced by $-\infty$, and $I(\varepsilon)$ taken to have the form given by (21.9). Then we have

$$\int_{-\infty}^{\infty} I(\varepsilon)\,\mathrm{d}\varepsilon = \frac{f^2\lambda}{4m}\int_{-\infty}^{\infty}\frac{\mathrm{d}\varepsilon}{\varepsilon^2+\lambda^2} = \frac{\pi f^2}{4m}. \tag{21.10}$$

§22. Parametric resonance

There exist oscillatory systems which are not closed, but in which the external action amounts only to a time variation of the parameters.†

The parameters of a one-dimensional system are the coefficients m and k in the Lagrangian (17.3). If these are functions of time, the equation of motion is

$$\frac{\mathrm{d}}{\mathrm{d}t}(m\dot{x})+kx = 0. \qquad (22.1)$$

We introduce instead of t a new independent variable τ such that $\mathrm{d}\tau = \mathrm{d}t/m(t)$; this reduces the equation to

$$\mathrm{d}^2x/\mathrm{d}\tau^2+mkx = 0.$$

There is therefore no loss of generality in considering an equation of motion of the form

$$\mathrm{d}^2x/\mathrm{d}t^2+\omega^2(t)x = 0 \qquad (22.2)$$

obtained from (22.1) if m = constant.

The form of the function $\omega(t)$ is given by the conditions of the problem. Let us assume that this function is periodic with some frequency γ and period $T = 2\pi/\gamma$. This means that $\omega(t+T) = \omega(t)$, and so the equation (22.2) is invariant under the transformation $t \to t+T$. Hence, if $x(t)$ is a solution of the equation, so is $x(t+T)$. That is, if $x_1(t)$ and $x_2(t)$ are two independent integrals of equation (22.2), they must be transformed into linear combinations of themselves when t is replaced by $t+T$. It is possible‡ to choose x_1 and x_2 in such a way that, when $t \to t+T$, they are simply multiplied by constants: $x_1(t+T) = \mu_1 x_1(t)$, $x_2(t+T) = \mu_2 x_2(t)$. The most general functions having this property are

$$x_1(t) = \mu_1^{t/T}\Pi_1(t), \qquad x_2(t) = \mu_2^{t/T}\Pi_2(t), \qquad (22.3)$$

where $\Pi_1(t)$, $\Pi_2(t)$ are purely periodic functions of time with period T.

---

† A simple example is that of a pendulum whose point of support executes a given periodic motion in a vertical direction (see Problem).

‡ Provided that the constants μ_1 and μ_2 are not equal.

The constants μ_1 and μ_2 in these functions must be related in a certain way. Multiplying the equations $\ddot{x}_1+\omega^2(t)x_1 = 0$, $\ddot{x}_2+\omega^2(t)x_2 = 0$ by x_2 and x_1 respectively and subtracting, we have $\ddot{x}_1x_2-\ddot{x}_2x_1 = d(\dot{x}_1x_2-x_1\dot{x}_2)/dt = 0$, or

$$\dot{x}_1x_2-x_1\dot{x}_2 = \text{constant}. \tag{22.4}$$

For any functions $x_1(t)$, $x_2(t)$ of the form (22.3), the expression on the left-hand side of (22.4) is multiplied by $\mu_1\mu_2$ when t is replaced by $t+T$. Hence it is clear that, if equation (22.4) is to hold, we must have

$$\mu_1\mu_2 = 1. \tag{22.5}$$

Further information about the constants μ_1, μ_2 can be obtained from the fact that the coefficients in equation (22.2) are real. If $x(t)$ is any integral of such an equation, then the complex conjugate function $x^*(t)$ must also be an integral. Hence it follows that μ_1, μ_2 must be the same as μ_1^*, μ_2^*, i.e. either $\mu_1 = \mu_2^*$ or μ_1 and μ_2 are both real. In the former case, (22.5) gives $\mu_1 = 1/\mu_1^*$, i.e. $|\mu_1|^2 = |\mu_2|^2 = 1$: the constants μ_1 and μ_2 are of modulus unity.

In the other case, two independent integrals of equation (22.2) are

$$x_1(t) = \mu^{t/T}\Pi_1(t), \qquad x_2(t) = \mu^{-t/T}\Pi_2(t), \tag{22.6}$$

with a positive or negative real value of μ ($|\mu| \neq 1$). One of these functions (x_1 or x_2 according as $|\mu| > 1$ or $|\mu| < 1$) increases exponentially with time. This means that the system at rest in equilibrium ($x = 0$) is unstable: any deviation from this state, however small, is sufficient to lead to a rapidly increasing displacement x. This is called *parametric resonance.*

It should be noticed that, when the initial values of x and $\dot{x}$ are exactly zero, they remain zero, unlike what happens in ordinary resonance (§18), in which the displacement increases with time (proportionally to t) even from initial values of zero.

Let us determine the conditions for parametric resonance to occur in the important case where the function $\omega(t)$ differs only slightly from a constant value ω_0 and is a simple periodic function:

$$\omega^2(t) = \omega_0^2(1+h\cos\gamma t), \tag{22.7}$$

where the constant $h \ll 1$; we shall suppose h positive, as may always be done by suitably choosing the origin of time. As we shall see below, parametric resonance is strongest if the frequency of the function $\omega(t)$ is nearly twice ω_0. Hence we put $\gamma = 2\omega_0+\varepsilon$, where $\varepsilon \ll \omega_0$.

The solution of the equation of motion

$$\ddot{x}+\omega_0^2[1+h\cos(2\omega_0+\varepsilon)t]x = 0 \tag{22.8}$$

may be sought in the form

$$x = a(t)\cos(\omega_0+\tfrac{1}{2}\varepsilon)t+b(t)\sin(\omega_0+\tfrac{1}{2}\varepsilon)t, \tag{22.9}$$

where $a(t)$ and $b(t)$ are functions of time which vary slowly in comparison with the trigonometrical factors. This form of solution is, of course, not exact. In reality, the function $x(t)$ also involves terms with frequencies which differ from $\omega_0+\frac{1}{2}\varepsilon$ by integral multiples of $2\omega_0+\varepsilon$; these terms are, however, of a higher order of smallness with respect to h, and may be neglected in a first approximation.

The values of γ which divide the regions of instability and stability correspond to $\mu = 1$ in (22.6) and to time-independent coefficients a and b in (22.9). Thus the determination of the limits of the resonance region amounts to finding the values of γ (or, equivalently, those of ε) for which the equation of motion is satisfied (with the necessary accuracy) by the solution (22.9) with constant a and b.

We substitute (22.9) in (22.8). The products of trigonometrical functions may be replaced by sums:

$$\cos(\omega_0+\tfrac{1}{2}\varepsilon)t\cos(2\omega_0+\varepsilon)t = \tfrac{1}{2}\cos 3(\omega_0+\tfrac{1}{2}\varepsilon)t+\tfrac{1}{2}\cos(\omega_0+\tfrac{1}{2}\varepsilon)t,$$

etc., and in accordance with what was said above we omit terms with frequency $3(\omega_0+\frac{1}{2}\varepsilon)$. The result is

$$b(\varepsilon+\tfrac{1}{2}h\omega_0)\sin(\omega_0+\tfrac{1}{2}\varepsilon)t+a(\varepsilon-\tfrac{1}{2}h\omega_0)\cos(\omega_0+\tfrac{1}{2}\varepsilon)t = 0.$$

If this equation is to be satisfied, the coefficients of the sine and cosine must both be zero. Thus either $\varepsilon = -\frac{1}{2}h\omega_0$ and $a = 0$, or $\varepsilon = \frac{1}{2}h\omega_0$ and $b = 0$. These values of ε give the limits of the region of parametric resonance. Thus parametric resonance occurs in the range

$$-\tfrac{1}{2}h\omega_0 < \varepsilon < \tfrac{1}{2}h\omega_0 \tag{22.10}$$

on either side of the frequency $2\omega_0$.

Parametric resonance also occurs when the frequency γ is close to any value $2\omega_0/n$ with n integral. The width of the resonance range decreases rapidly with increasing n, however, namely as h^n.

PROBLEM

Find the conditions for parametric resonance in small oscillations of a simple pendulum whose point of support oscillates vertically.

SOLUTION. The Lagrangian derived in §5, Problem 2, gives for small oscillations ($\phi \ll 1$) the equation of motion $\ddot{\phi}+\omega_0^2[1+(4a/l)\cos(2\omega_0+\epsilon)t]\phi = 0$, where $\omega_0^2 = g/l$. Hence we see that the parameter h is here represented by $4a/l$. The condition (22.10) becomes $|\epsilon| < 2a\sqrt{(g/l^3)}$.

§23. Anharmonic oscillations

The whole of the theory of small oscillations discussed above is based on the expansion of the potential and kinetic energies of the system in terms of the coordinates and velocities, retaining only the second-order terms. The equations of motion are then linear, and in this approximation we speak of *linear oscillations*. Although such an expansion is entirely legitimate when the amplitude of the oscillations is sufficiently small, in higher approximations (called *anharmonic* or *non-linear oscillations*) some minor but qualitatively different properties of the motion appear.

Let us consider the expansion of the Lagrangian as far as the third-order terms. In the potential energy there appear terms of degree three in the coordinates x_i, and in the kinetic energy terms containing products of velocities and coordinates, of the form $\dot{x}_i\dot{x}_k x_l$. This difference from the previous expression (19.3) is due to the retention of terms linear in x in the expansion of the functions $a_{ik}(q)$. Thus the Lagrangian is of the form

$$L = \tfrac{1}{2}\sum_{i,k}(m_{ik}\dot{x}_i\dot{x}_k - k_{ik}x_i x_k) + \\ +\tfrac{1}{2}\sum_{i,k,l} n_{ikl}\dot{x}_i\dot{x}_k x_l - \tfrac{1}{3}\sum_{i,k,l} l_{ikl}x_i x_k x_l, \qquad (23.1)$$

where n_{ikl}, l_{ikl} are further constant coefficients.

If we change from arbitrary coordinates x_i to the normal coordinates Q_α of the linear approximation, then, because this transformation is linear, the third and fourth sums in (23.1) become similar sums with Q_α and $\dot{Q}_\alpha$ in place of the coordinates x_i and the velocities $\dot{x}_i$. Denoting the coefficients in these new sums by $\lambda_{\alpha\beta\gamma}$ and $\mu_{\alpha\beta\gamma}$, we have the Lagrangian in the form

$$L = \tfrac{1}{2}\sum_{\alpha}(\dot{Q}_\alpha^2 - \omega_\alpha^2 Q_\alpha^2) + \tfrac{1}{2}\sum_{\alpha,\beta,\gamma}\lambda_{\alpha\beta\gamma}\dot{Q}_\alpha\dot{Q}_\beta Q_\gamma - \tfrac{1}{3}\sum_{\alpha,\beta,\gamma}\mu_{\alpha\beta\gamma}Q_\alpha Q_\beta Q_\gamma. \tag{23.2}$$

We shall not pause to write out in their entirety the equations of motion derived from this Lagrangian. The important feature of these equations is that they are of the form

$$\ddot{Q}_\alpha + \omega_\alpha^2 Q_\alpha = f_\alpha(Q, \dot{Q}, \ddot{Q}), \tag{23.3}$$

where f_α are homogeneous functions, of degree two, of the coordinates Q and their time derivatives.

Using the method of successive approximations, we seek a solution of these equations in the form

$$Q_\alpha = Q_\alpha^{(1)} + Q_\alpha^{(2)}, \tag{23.4}$$

where $Q_\alpha^{(2)} \ll Q_\alpha^{(1)}$, and the $Q_\alpha^{(1)}$ satisfy the "unperturbed" equations $\ddot{Q}_\alpha^{(1)} + \omega_\alpha^2 Q_\alpha^{(1)} = 0$, i.e. they are ordinary harmonic oscillations:

$$Q_\alpha^{(1)} = a_\alpha \cos(\omega_\alpha t + \alpha_\alpha). \tag{23.5}$$

Retaining only the second-order terms on the right-hand side of (23.3) in the next approximation, we have for the $Q_\alpha^{(2)}$ the equations

$$\ddot{Q}_\alpha^{(2)} + \omega_\alpha^2 Q_\alpha^{(2)} = f_\alpha(Q^{(1)}, \dot{Q}^{(1)}, \ddot{Q}^{(1)}), \tag{23.6}$$

where (23.5) is to be substituted on the right. This gives a set of inhomogeneous linear differential equations, in which the right-hand sides can be represented as sums of simple periodic functions. For example,

$$\begin{aligned} Q_\alpha^{(1)} Q_\beta^{(1)} &= a_\alpha a_\beta \cos(\omega_\alpha t + \alpha_\alpha)\cos(\omega_\beta t + \alpha_\beta) \\ &= \tfrac{1}{2} a_\alpha a_\beta \{\cos[(\omega_\alpha + \omega_\beta)t + \alpha_\alpha + \alpha_\beta] + \cos[(\omega_\alpha - \omega_\beta)t + \alpha_\alpha - \alpha_\beta]\}. \end{aligned}$$

Thus the right-hand sides of equations (23.6) contain terms corresponding to oscillations whose frequencies are the sums and differences of the eigenfrequencies of the system. The solution of these equations must be sought in a form involving similar periodic factors, and so we conclude that, in the second approximation, additional oscillations with frequencies

$$\omega_\alpha \pm \omega_\beta, \tag{23.7}$$

including the double frequencies $2\omega_\alpha$ and the frequency zero (corresponding to a constant displacement), are superposed on the normal oscillations of the system. These are called *combination frequencies*. The corresponding amplitudes are proportional to the products $a_\alpha a_\beta$ (or the squares a_α^2) of the corresponding normal amplitudes.

In higher approximations, when further terms are included in the expansion of the Lagrangian, combination frequencies occur which are the sums and differences of more than two ω_α; and a further phenomenon also appears. In the third approximation, the combination frequencies include some which coincide with the original frequencies $\omega_\alpha (= \omega_\alpha + \omega_\beta - \omega_\beta)$. When the method described above is used, the right-hand sides of the equations of motion therefore include resonance terms, which lead to terms in the solution whose amplitude increases with time. It is physically evident, however, that the magnitude of the oscillations cannot increase of itself in a closed system with no external source of energy.

In reality, the fundamental frequencies ω_α in higher approximations are not equal to their "unperturbed" values $\omega_\alpha^{(0)}$ which appear in the quadratic expression for the potential energy. The increasing terms in the solution arise from an expansion of the type

$$\cos(\omega_\alpha^{(0)} + \Delta\omega_\alpha)t \approx \cos\omega_\alpha^{(0)}t - t\Delta\omega_\alpha \sin\omega_\alpha^{(0)}t,$$

which is obviously not legitimate when t is sufficiently large.

CHAPTER 6

MOTION OF A RIGID BODY

§24. Angular velocity

A *rigid body* may be defined in mechanics as a system of particles such that the distances between the particles do not vary. This condition can, of course, be satisfied only approximately by systems which actually exist in nature. The majority of solid bodies, however, change so little in shape and size under ordinary conditions that these changes may be entirely neglected in considering the laws of motion of the body as a whole.

In what follows, we shall often simplify the derivations by regarding a rigid body as a discrete set of particles, but this in no way invalidates the assertion that solid bodies may usually be regarded in mechanics as continuous, and their internal structure disregarded. The passage from the formulae which involve a summation over discrete particles to those for a continuous body is effected by simply replacing the mass of each particle by the mass $\varrho\,\mathrm{d}V$ contained in a volume element $\mathrm{d}V$ (ϱ being the density) and the summation by an integration over the volume of the body.

To describe the motion of a rigid body, we use two systems of coordinates: a "fixed" (i.e. inertial) system XYZ, and a moving system $x_1 = x$, $x_2 = y$, $x_3 = z$ which is supposed to be rigidly fixed in the body and to participate in its motion. The origin of the moving system may conveniently be taken to coincide with the centre of mass of the body.

The position of the body with respect to the fixed system of coordinates is completely determined if the position of the moving system is specified. Let the origin O of the moving system have the position $\mathbf{R}_0$ (Fig. 19). The orientation of the axes of that system relative to the fixed system is given by three independent angles, which together with the three components of the vector $\mathbf{R}_0$ make six coordinates. Thus a rigid body is a mechanical system with six degrees of freedom.

Let us consider an arbitrary infinitesimal displacement of a rigid body. It can be represented as the sum of two parts. One of these is an infinitesimal translation of the body, whereby the centre of mass moves to its final position, but the orientation of the axes of the moving system of coordinates is unchanged. The other is an infinitesimal rotation about the centre of mass, whereby the remainder of the body moves to its final position.

Let $\mathbf{r}$ be the position of an arbitrary point P in a rigid body in the moving system, and $\mathbf{R}$ the position of the same point in the fixed system (Fig. 19). Then the infinitesimal displacement $\mathrm{d}\mathbf{R}$ of P consists of a displacement $\mathrm{d}\mathbf{R}_0$, equal to that of the centre of mass, and a displacement $\mathrm{d}\boldsymbol{\phi}\times\mathbf{r}$ relative to the centre of mass resulting from a rotation through an infinitesimal angle $\mathrm{d}\phi$ (see (9.1)): $\mathrm{d}\mathbf{R} = \mathrm{d}\mathbf{R}_0+\mathrm{d}\boldsymbol{\phi}\times\mathbf{r}$. Dividing this equation by the time $\mathrm{d}t$ during which the displacement occurs, and putting

$$\mathrm{d}\mathbf{R}/\mathrm{d}t = \mathbf{v}, \qquad \mathrm{d}\mathbf{R}_0/\mathrm{d}t = \mathbf{V}, \qquad \mathrm{d}\boldsymbol{\phi}/\mathrm{d}t = \boldsymbol{\Omega}, \tag{24.1}$$

we obtain the relation

$$\mathbf{v} = \mathbf{V}+\boldsymbol{\Omega}\times\mathbf{r}. \tag{24.2}$$

The vector $\mathbf{V}$ is the velocity of the centre of mass of the body, and is also the *translational velocity* of the body. The vector $\boldsymbol{\Omega}$ is called the *angular velocity* of the rotation of the body; its direction, like that of $\mathrm{d}\boldsymbol{\phi}$, is along the axis of rotation. Thus the velocity $\mathbf{v}$ of any point in the body relative to the fixed system of coordinates can be expressed in terms of the translational velocity of the body and its angular velocity of rotation.

It should be emphasised that, in deriving formula (24.2), no use has been made of the fact that the origin is located at the centre of

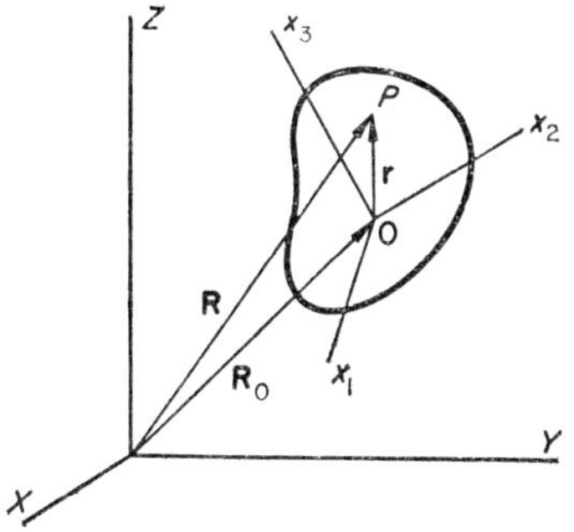

FIG. 19

mass. The advantages of this choice of origin will become evident when we come to calculate the energy of the moving body.

Let us now assume that the system of coordinates fixed in the body is such that its origin is not at the centre of mass O, but at some point O' at a distance $\mathbf{a}$ from O. Let the velocity of O' be $\mathbf{V}'$, and the angular velocity of the new system of coordinates be $\mathbf{\Omega}'$. We again consider some point P in the body, and denote by $\mathbf{r}'$ its position with respect to O'. Then $\mathbf{r} = \mathbf{r}'+\mathbf{a}$, and substitution in (24.2) gives $\mathbf{v} = \mathbf{V}+\mathbf{\Omega}\times\mathbf{a}+\mathbf{\Omega}\times\mathbf{r}'$. The definition of $\mathbf{V}'$ and $\mathbf{\Omega}'$ shows that $\mathbf{v} = \mathbf{V}'+\mathbf{\Omega}'\times\mathbf{r}'$. Hence it follows that

$$\mathbf{V}' = \mathbf{V}+\mathbf{\Omega}\times\mathbf{a}, \qquad \mathbf{\Omega}' = \mathbf{\Omega}. \tag{24.3}$$

The second of these equations is very important. We see that the angular velocity of rotation, at any instant, of a system of coordinates fixed in the body is independent of the particular system chosen. All such systems rotate with angular velocities $\mathbf{\Omega}$ which are equal in magnitude and parallel in direction. This enables us to call $\mathbf{\Omega}$ the *angular velocity of the body*. The velocity of the translational motion, however, does not have this "absolute" property.

It is seen from the first formula (24.3) that, if $\mathbf{V}$ and $\mathbf{\Omega}$ are, at any given instant, perpendicular for some choice of the origin O, then $\mathbf{V}'$ and $\mathbf{\Omega}'$ are perpendicular for any other origin O'. Formula (24.2) shows that in this case the velocities $\mathbf{v}$ of all points in the body are in a plane perpendicular to $\mathbf{\Omega}$. It is then always possible† to choose an origin O' whose velocity $\mathbf{V}'$ is zero, so that the motion of the body at the instant considered is a pure rotation about an axis through O'. This axis is called the *instantaneous axis of rotation*.‡

In what follows we shall always suppose that the origin of the moving system is taken to be at the centre of mass of the body, and so the axis of rotation passes through the centre of mass. In general both the magnitude and the direction of $\mathbf{\Omega}$ vary during the motion.

§25. The inertia tensor

To calculate the kinetic energy of a rigid body, we may consider it as a discrete system of particles and put $T = \Sigma\frac{1}{2}mv^2$, where the summation is taken over all the particles in the body. Here, and in what follows, we simplify the notation by omitting the suffix which denumerates the particles.

Substitution of (24.2) gives

$$T = \sum \tfrac{1}{2}m(\mathbf{V}+\mathbf{\Omega}\times\mathbf{r})^2 = \sum \tfrac{1}{2}mV^2+\sum m\mathbf{V}\cdot\mathbf{\Omega}\times\mathbf{r}+\sum \tfrac{1}{2}m(\mathbf{\Omega}\times\mathbf{r})^2.$$

The velocities $\mathbf{V}$ and $\mathbf{\Omega}$ are the same for every point in the body. In the first term, therefore, $\frac{1}{2}V^2$ can be taken outside the summation sign, and Σm is just the mass of the body, which we denote by μ. In the second term we put $\Sigma m\mathbf{V}\cdot\mathbf{\Omega}\times\mathbf{r} = \Sigma m\mathbf{r}\cdot\mathbf{V}\times\mathbf{\Omega} = \mathbf{V}\times\mathbf{\Omega}\cdot\Sigma m\mathbf{r}$. Since we take the origin of the moving system to be at the centre of mass, this term is zero, because $\Sigma m\mathbf{r} = 0$. Finally, in the third term we expand the squared vector product. The result is

$$T = \tfrac{1}{2}\mu V^2+\tfrac{1}{2}\sum m[\Omega^2r^2-(\mathbf{\Omega}\cdot\mathbf{r})^2]. \tag{25.1}$$

Thus the kinetic energy of a rigid body can be written as the sum of two parts. The first term in (25.1) is the kinetic energy of the translation-

† O' may, of course, lie outside the body.

‡ In the general case where $\mathbf{V}$ and $\mathbf{\Omega}$ are not perpendicular, the origin may be chosen so as to make $\mathbf{V}$ and $\mathbf{\Omega}$ parallel, i.e. so that the motion consists (at the instant in question) of a rotation about some axis together with a translation along that axis.

al motion, and is of the same form as if the whole mass of the body were concentrated at the centre of mass. The second term is the kinetic energy of the rotation with angular velocity $\mathbf{\Omega}$ about an axis passing through the centre of mass. It should be emphasised that this division of the kinetic energy into two parts is possible only because the origin of the coordinate system fixed in the body has been taken to be at its centre of mass.

We may rewrite the kinetic energy of rotation in tensor form, i.e. in terms of the components† x_i and Ω_i of the vectors $\mathbf{r}$ and $\mathbf{\Omega}$. We have

$$\begin{aligned} T_{\text{rot}} &= \tfrac{1}{2}\sum m(\Omega_i^2 x_l^2 - \Omega_i x_i \Omega_k x_k) \\ &= \tfrac{1}{2}\sum m(\Omega_i \Omega_k \delta_{ik} x_l^2 - \Omega_i \Omega_k x_i x_k) \\ &= \tfrac{1}{2}\Omega_i \Omega_k \sum m(x_l^2 \delta_{ik} - x_i x_k). \end{aligned}$$

Here we have used the identity $\Omega_i = \delta_{ik}\Omega_k$, where δ_{ik} is the unit tensor, whose components are unity for $i = k$ and zero for $i \neq k$. In terms of the tensor

$$I_{ik} = \sum m(x_l^2 \delta_{ik} - x_i x_k) \tag{25.2}$$

we have finally the following expression for the kinetic energy of a rigid body:

$$T = \tfrac{1}{2}\mu V^2 + \tfrac{1}{2}I_{ik}\Omega_i\Omega_k. \tag{25.3}$$

The Lagrangian for a rigid body is obtained from (25.3) by subtracting the potential energy:

$$L = \tfrac{1}{2}\mu V^2 + \tfrac{1}{2}I_{ik}\Omega_i\Omega_k - U. \tag{25.4}$$

The potential energy is in general a function of the six variables which define the position of the rigid body, e.g. the three coordinates X, Y, Z of the centre of mass and the three angles which specify the relative orientation of the moving and fixed coordinate axes.

---

† The letters i, k, l are tensor suffixes and take the values 1, 2, 3. The summation rule will always be used, i.e. summation signs are omitted, but summation over the values 1, 2, 3 is implied whenever a suffix occurs twice in any expression. Such a suffix is called a *dummy suffix*. For example, $A_i B_i = \mathbf{A}\cdot\mathbf{B}$, $A_i^2 = A_i A_i = \mathbf{A}^2$, etc. It is obvious that dummy suffixes can be replaced by any other like suffixes, except ones which already appear elsewhere in the expression concerned.

The tensor I_{ik} is called the *inertia tensor* of the body. It is symmetrical, i.e.

$$I_{ik} = I_{ki}, \tag{25.5}$$

as is evident from the definition (25.2). For clarity, we may give its components explicitly:

$$I_{ik} = \begin{bmatrix} \sum m(y^2+z^2) & -\sum mxy & -\sum mxz \\ -\sum myx & \sum m(x^2+z^2) & -\sum myz \\ -\sum mzx & -\sum mzy & \sum m(x^2+y^2) \end{bmatrix}. \tag{25.6}$$

The components I_{xx}, I_{yy}, I_{zz} are called the *moments of inertia* about the corresponding axes.

The inertia tensor is evidently additive: the moments of inertia of a body are the sums of those of its parts.

If the body is regarded as continuous, with density ϱ, the sum in the definition (25.2) becomes an integral over the volume of the body:

$$I_{ik} = \int \varrho(x_l^2\delta_{ik} - x_i x_k)\, \mathrm{d}V. \tag{25.7}$$

Like any symmetrical tensor of rank two, the inertia tensor can be reduced to diagonal form by an appropriate choice of the directions of the axes x_1, x_2, x_3. These directions are called the *principal axes of inertia*, and the corresponding values of the diagonal components of the tensor are called the *principal moments of inertia*; we shall denote them by I_1, I_2, I_3. When the axes x_1, x_2, x_3 are so chosen, the kinetic energy of rotation takes the very simple form

$$T_{\text{rot}} = \tfrac{1}{2}(I_1\Omega_1^2 + I_2\Omega_2^2 + I_3\Omega_3^2). \tag{25.8}$$

None of the three principal moments of inertia can exceed the sum of the other two. For instance,

$$I_1 + I_2 = \sum m(x_1^2 + x_2^2 + 2x_3^2) \geqslant \sum m(x_1^2 + x_2^2) = I_3. \tag{25.9}$$

A body whose three principal moments of inertia are all different is called an *asymmetrical top*. If two are equal ($I_1 = I_2 \neq I_3$), we have a *symmetrical top*. In this case the direction of one of the principal axes in the x_1x_2 plane may be chosen arbitrarily. If all three principal moments of inertia are equal, the body is called a *spherical top*, and the three axes of inertia may be chosen arbitrarily as any three mutually perpendicular axes.

The determination of the principal axes of inertia is much simplified if the body is symmetrical, for it is clear that the position of the centre of mass and the directions of the principal axes must have the same symmetry as the body. For example, if the body has a plane of symmetry, the centre of mass must lie in that plane, which also contains two of the principal axes of inertia, while the third is perpendicular to the plane. An obvious case of this kind is a coplanar system of particles. Here there is a simple relation between the three principal moments of inertia. If the plane of the system is taken as the x_1x_2 plane, then $x_3 = 0$ for every particle, and so $I_1 = \Sigma mx_2^2$, $I_2 = \Sigma mx_1^2$, $I_3 = \Sigma m(x_1^2+x_2^2)$, whence

$$I_3 = I_1+I_2. \tag{25.10}$$

If a body has an axis of symmetry of any order, the centre of mass must lie on that axis, which is also one of the principal axes of inertia, while the other two are perpendicular to it. If the axis is of order higher than the second, the body is a symmetrical top. For any principal axis perpendicular to the axis of symmetry can be turned through an angle different from 180° about the latter, i.e. the choice of the perpendicular axes is not unique, and this can happen only if the body is a symmetrical top.

A particular case here is a collinear system of particles. If the line of the system is taken as the x_3 axis, then $x_1 = x_2 = 0$ for every particle, and so two of the principal moments of inertia are equal and the third is zero:

$$I_1 = I_2 = \sum mx_3^2, \qquad I_3 = 0. \tag{25.11}$$

Such a system is called a *rotator*. The characteristic property which distinguishes a rotator from other bodies is that it has only two, not three, rotational degrees of freedom, corresponding to rotations about the x_1 and x_2 axes: it is clearly meaningless to speak of the rotation of a straight line about itself.

Finally, we may note one further result concerning the calculation of the inertia tensor. Although this tensor has been defined with respect to a system of coordinates whose origin is at the centre of mass (as is necessary if the fundamental formula (25.3) is to be valid), it may sometimes be more conveniently found by first calculating a similar tensor $I'_{ik} = \Sigma m(x_l'^2\delta_{ik} - x_i'x_k')$, defined with respect to some other origin O'. If the distance OO' is represented by a vector $\mathbf{a}$, then $\mathbf{r} = \mathbf{r}'+\mathbf{a}$, $x_i = x_i'+a_i$; since, by the definition of O, $\Sigma m\mathbf{r} = 0$, we have

$$I'_{ik} = I_{ik} + \mu(a^2\delta_{ik} - a_i a_k). \tag{25.12}$$

Using this formula, we can easily calculate I_{ik} if I'_{ik} is known.

PROBLEMS

PROBLEM 1. Determine the principal moments of inertia for the following types of molecule, regarded as systems of particles at fixed distances apart: (a) a molecule of three collinear atoms, (b) a triatomic molecule which is an isosceles triangle (Fig. 20).

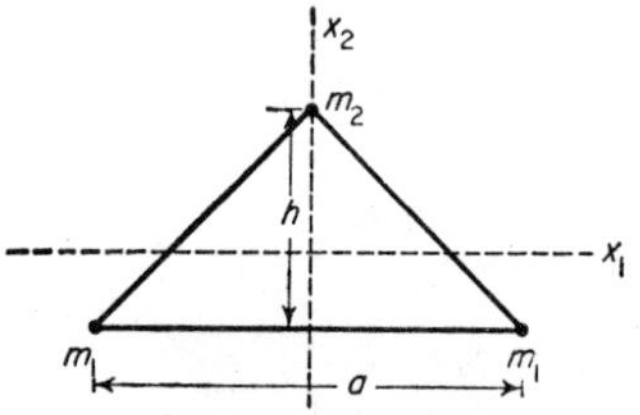

FIG. 20

SOLUTION. (a)

$$I_1 = I_2 = \frac{1}{\mu}(m_1m_2l_{12}^2 + m_1m_3l_{13}^2 + m_2m_3l_{23}^2), \qquad I_3 = 0,$$

where m_a is the mass of the ath atom and l_{ab} the distance between the ath and bth atoms.

For a diatomic molecule the result is obvious: it is the product of the reduced mass of the two atoms and the square of the distance between them: $I_1 = I_2 = m_1m_2l^2/(m_1+m_2)$.

(b) The centre of mass is on the axis of symmetry of the triangle, at a distance m_2h/μ from its base (h being the height of the triangle). The moments of inertia are $I_1 = 2m_1m_2h^2/\mu$, $I_2 = \frac{1}{2}m_1a^2$, $I_3 = I_1+I_2$.

PROBLEM 2. Determine the principal moments of inertia for the following homogeneous bodies: (a) a thin rod of length l, (b) a sphere of radius R, (c) a circular cylinder of radius R and height h, (d) a rectangular parallelepiped of sides a, b, and c, (e) an ellipsoid of semiaxes a, b, c.

SOLUTION. (a) $I_1 = I_2 = \frac{1}{12}\mu l^2$, $I_3 = 0$ (we neglect the thickness of the rod).

(b) $I_1 = I_2 = I_3 = \frac{2}{5}\mu R^2$ (found by calculating the sum $I_1+I_2+I_3 = 2\varrho \int r^2\,\mathrm{d}V$).

(c) $I_1 = I_2 = \frac{1}{4}\mu(R^2+\frac{1}{3}h^2)$, $I_3 = \frac{1}{2}\mu R^2$ (where the x_3 axis is along the axis of the cylinder).

(d) $I_1 = \frac{1}{12}\mu(b^2+c^2)$, $I_2 = \frac{1}{12}\mu(a^2+c^2)$, $I_3 = \frac{1}{12}\mu(a^2+b^2)$ (where the axes x_1, x_2, x_3 are along the sides a, b, c respectively).

(e) The centre of mass is at the centre of the ellipsoid, and the principal axes of inertia are along the axes of the ellipsoid. The integration over the volume of the ellipsoid can be reduced to one over a sphere by the transformation $x = a\xi$, $y = b\eta$, $z = c\zeta$, which converts the equation of the surface of the ellipsoid $x^2/a^2+y^2/b^2+z^2/c^2 = 1$ into that of the unit sphere $\xi^2+\eta^2+\zeta^2 = 1$.

For example, the moment of inertia about the x axis is

$$\begin{aligned} I_1 &= \varrho \iiint (y^2+z^2)\,\mathrm{d}x\,\mathrm{d}y\,\mathrm{d}z \\ &= \varrho\, abc \iiint (b^2\eta^2+c^2\zeta^2)\,\mathrm{d}\xi\,\mathrm{d}\eta\,\mathrm{d}\zeta \\ &= \tfrac{1}{2}abcI'(b^2+c^2), \end{aligned}$$

where I' is the moment of inertia of a sphere of unit radius. Since the volume of the ellipsoid is $4\pi abc/3$, we find the moments of inertia $I_1 = \frac{1}{5}\mu(b^2+c^2)$, $I_2 = \frac{1}{5}\mu(a^2+c^2)$, $I_3 = \frac{1}{5}\mu(a^2+b^2)$.

PROBLEM 3. Determine the frequency of small oscillations of a compound pendulum (a rigid body swinging about a fixed horizontal axis in a gravitational field).

SOLUTION. Let l be the distance between the centre of mass of the pendulum and the axis about which it rotates, and α, β, γ the angles between the principal axes of inertia and the axis of rotation. We take as the variable coordinate the angle ϕ between the vertical and a line through the centre of mass perpendicular to the axis of rotation. The velocity of the centre of mass is $V = l\dot{\phi}$, and the components of the angular velocity along the principal axes of inertia are $\dot{\phi}\cos\alpha$, $\dot{\phi}\cos\beta$, $\dot{\phi}\cos\gamma$. Assuming the angle ϕ to be small, we find the potential energy $U = \mu gl(1-\cos\phi) \approx \frac{1}{2}\mu gl\phi^2$. The Lagrangian is therefore

$$L = \tfrac{1}{2}\mu l^2\dot{\phi}^2+\tfrac{1}{2}(I_1\cos^2\alpha+I_2\cos^2\beta+I_3\cos^2\gamma)\dot{\phi}^2-\tfrac{1}{2}\mu gl\phi^2.$$

The frequency of the oscillations is consequently

$$\omega^2 = \mu gl/(\mu l^2+I_1\cos^2\alpha+I_2\cos^2\beta+I_3\cos^2\gamma).$$

PROBLEM 4. Find the kinetic energy of the system shown in Fig. 21: OA and AB are thin uniform rods of length l hinged together at A. The rod OA rotates (in the plane of the diagram) about O, while the end B of the rod AB slides along Ox.

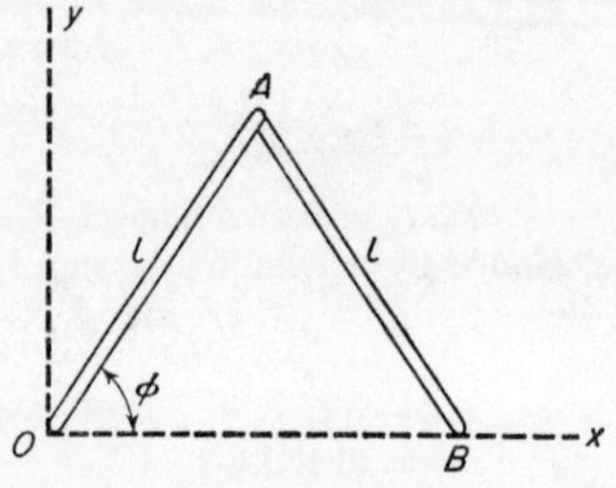

FIG. 21

SOLUTION. The velocity of the centre of mass of the rod OA (which is at the middle of the rod) is $\frac{1}{2}l\dot{\phi}$, where ϕ is the angle AOB. The kinetic energy of the rod OA is therefore $T_1 = \frac{1}{8}\mu l^2\dot{\phi}^2 + \frac{1}{2}I\dot{\phi}^2$, where μ is the mass of each rod.

The Cartesian coordinates of the centre of mass of the rod AB are $X = \frac{3}{2}l\cos\phi$, $Y = \frac{1}{2}l\sin\phi$. Since the angular velocity of rotation of this rod is also $\dot{\phi}$, its kinetic energy is $T_2 = \frac{1}{2}\mu(\dot{X}^2+\dot{Y}^2)+\frac{1}{2}I\dot{\phi}^2 = \frac{1}{8}\mu l^2(1+8\sin^2\phi)\dot{\phi}^2+\frac{1}{2}I\dot{\phi}^2$. The total kinetic energy of this system is therefore $T = \frac{1}{3}\mu l^2(1+3\sin^2\phi)\dot{\phi}^2$, since $I = \frac{1}{12}\mu l^2$ (see Problem 2(a)).

PROBLEM 5. Find the kinetic energy of a cylinder of radius R rolling on a plane, if the mass of the cylinder is so distributed that one of the principal axes of inertia is parallel to the axis of the cylinder and at a distance a from it, and the moment of inertia about that principal axis is I.

SOLUTION. Let ϕ be the angle between the vertical and a line from the centre of mass perpendicular to the axis of the cylinder (Fig. 22). The motion of the cylinder at any instant may be regarded as a pure rotation about an instantaneous axis which coincides with the line where the cylinder touches the plane. The angular velocity of this rotation is $\dot{\phi}$, since the angular velocity of rotation about all parallel axes is the same. The centre of mass is at a distance $\sqrt{(a^2+R^2-2aR\cos\phi)}$ from the instanta-

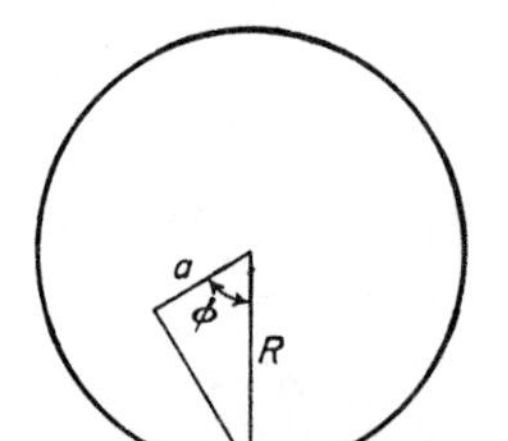

FIG. 22

neous axis, and its velocity is therefore $V = \dot{\phi}\sqrt{(a^2+R^2-2aR\cos\phi)}$. The total kinetic energy is

$$T = \tfrac{1}{2}\mu(a^2+R^2-2aR\cos\phi)\dot{\phi}^2+\tfrac{1}{2}I\dot{\phi}^2.$$

PROBLEM 6. Find the kinetic energy of a homogeneous cylinder of radius a rolling inside a cylindrical surface of radius R (Fig. 23).

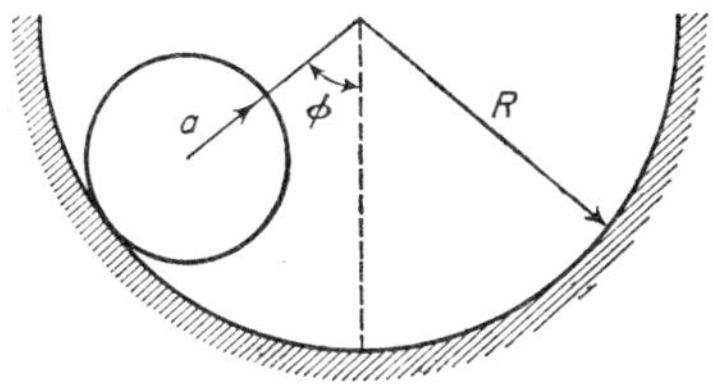

FIG. 23

SOLUTION. We use the angle ϕ between the vertical and the line joining the centres of the cylinders. The centre of mass of the rolling cylinder is on the axis, and its velocity is $V = \dot{\phi}(R-a)$. We can calculate the angular velocity as that of a pure rotation about an instantaneous axis which coincides with the line of contact of the cylinders; it is $\Omega = V/a = \dot{\phi}(R-a)/a$. If I_3 is the moment of inertia about the axis of the cylinder, then

$$T = \tfrac{1}{2}\mu(R-a)^2\dot{\phi}^2+\tfrac{1}{2}I_3(R-a)^2\dot{\phi}^2/a^2 = \tfrac{3}{4}\mu(R-a)^2\dot{\phi}^2,$$

I_3 being given by Problem 2(c).

§26. Angular momentum of a rigid body

The value of the angular momentum of a system depends, as we know, on the point with respect to which it is defined. In the mechanics of a rigid body, the most appropriate point to choose for this purpose is the origin of the moving system of coordinates, i.e. the centre of mass of the body, and in what follows we shall denote by **M** the angular momentum so defined.

According to formula (9.6), when the origin is taken at the centre of mass of the body, the angular momentum **M** is equal to the "intrinsic" angular momentum resulting from the motion relative to the centre of mass. In the definition $\mathbf{M} = \Sigma m\mathbf{r}\times\mathbf{v}$ we therefore replace

v by $\mathbf{\Omega}\times\mathbf{r}$:

$$\mathbf{M} = \sum m\mathbf{r}\times(\mathbf{\Omega}\times\mathbf{r}) = \sum m[r^2\mathbf{\Omega} - \mathbf{r}(\mathbf{r}\cdot\mathbf{\Omega})],$$

or, in tensor notation,

$$M_i = \sum m(x_l^2\Omega_i - x_i x_k \Omega_k) = \Omega_k \sum m(x_l^2\delta_{ik} - x_i x_k).$$

Finally, using the definition (25.2) of the inertia tensor, we have

$$M_i = I_{ik}\Omega_k. \tag{26.1}$$

If the axes x_1, x_2, x_3 are the same as the principal axes of inertia, formula (26.1) gives

$$M_1 = I_1\Omega_1, \quad M_2 = I_2\Omega_2, \quad M_3 = I_3\Omega_3. \tag{26.2}$$

In particular, for a spherical top, where all three principal moments of inertia are equal, we have simply

$$\mathbf{M} = I\mathbf{\Omega}, \tag{26.3}$$

i.e. the angular momentum vector is proportional to, and in the same direction as, the angular velocity vector. For an arbitrary body, however, the vector **M** is not in general in the same direction as $\mathbf{\Omega}$; this happens only when the body is rotating about one of its principal axes of inertia.

Let us consider a rigid body moving freely, i.e. not subject to any external forces. We suppose that any uniform translational motion, which is of no interest, is removed, leaving a free rotation of the body.

As in any closed system, the angular momentum of the freely rotating body is constant. For a spherical top the condition $\mathbf{M}$ = constant gives $\mathbf{\Omega}$ = constant; that is, the most general free rotation of a spherical top is a uniform rotation about an axis fixed in space.

The case of a rotator is equally simple. Here also $\mathbf{M} = I\mathbf{\Omega}$, and the vector $\mathbf{\Omega}$ is perpendicular to the axis of the rotator. Hence a free rotation of a rotator is a uniform rotation in one plane about an axis perpendicular to that plane.

The law of conservation of angular momentum also suffices to determine the more complex free rotation of a symmetrical top. Using the fact that the principal axes of inertia x_1, x_2 (perpendicular to the axis of symmetry (x_3) of the top) may be chosen arbitrarily, we take the x_2 axis perpendicular to the plane containing the constant vector **M** and the instantaneous position of the x_3 axis. Then $M_2 = 0$, and formulae (26.2) show that $\Omega_2 = 0$. This means that the directions of **M**, **Ω** and the axis of the top are at every instant in one plane (Fig. 24).

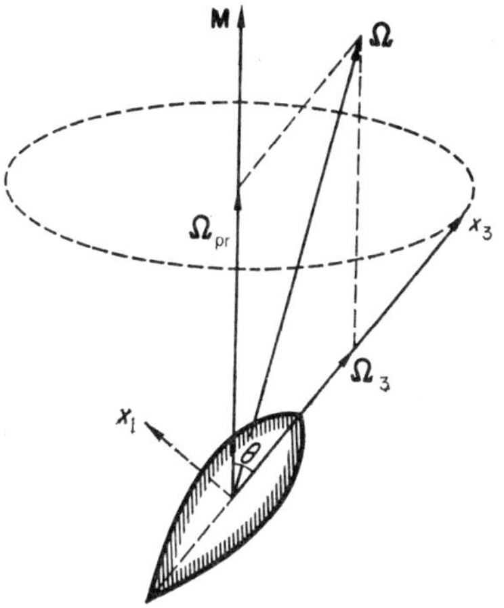

FIG. 24

Hence, in turn, it follows that the velocity $\mathbf{v} = \boldsymbol{\Omega} \times \mathbf{r}$ of every point on the axis of the top is at every instant perpendicular to that plane. That is, the axis of the top rotates uniformly (see below) about the direction of **M**, describing a circular cone. This is called *regular precession* of the top. At the same time the top rotates uniformly about its own axis.

The angular velocities of these two rotations can easily be expressed in terms of the given angular momentum M and the angle θ between the axis of the top and the direction of **M**. The angular velocity of the top about its own axis is just the component Ω_3 of the vector **Ω** along the axis:

$$\Omega_3 = M_3/I_3 = (M/I_3)\cos\theta. \tag{26.4}$$

To determine the rate of precession Ω_{pr}, the vector **Ω** must be resolved into components along x_3 and along **M**. The first of these gives no displacement of the axis of the top, and the second component is therefore the required angular velocity of precession. Fig. 24 shows that $\Omega_{pr} \sin\theta = \Omega_1$, and, since $\Omega_1 = M_1/I_1 = (M/I_1)\sin\theta$, we have

$$\Omega_{pr} = M/I_1. \tag{26.5}$$

§27. The equations of motion of a rigid body

Since a rigid body has, in general, six degrees of freedom, the general equations of motion must be six in number. They can be put in a form which gives the time derivatives of two vectors, the momentum and the angular momentum of the body.

The first equation is obtained by simply summing the equations $\dot{\mathbf{p}} = \mathbf{f}$ for each particle in the body, $\mathbf{p}$ being the momentum of the particle and $\mathbf{f}$ the force acting on it. In terms of the total momentum of the body $\mathbf{P} = \Sigma\mathbf{p} = \mu\mathbf{V}$ and total force acting on it $\mathbf{F} = \Sigma\mathbf{f}$, we have

$$\mathrm{d}\mathbf{P}/\mathrm{d}t = \mathbf{F}. \tag{27.1}$$

Although $\mathbf{F}$ has been defined as the sum of all the forces $\mathbf{f}$ acting on the various particles, including the forces due to other particles, $\mathbf{F}$ actually includes only external forces: the forces of interaction between the particles composing the body must cancel out, since if there are no external forces the momentum of the body, like that of any closed system, must be conserved, i.e. we must have $\mathbf{F} = 0$.

If U is the potential energy of a rigid body in an external field, the force $\mathbf{F}$ is obtained by differentiating U with respect to the coordinates of the centre of mass of the body:

$$\mathbf{F} = -\partial U/\partial \mathbf{R}_0. \tag{27.2}$$

For, when the body undergoes a translation through a distance $\delta\mathbf{R}_0$, the position vector $\mathbf{R}$ of every point in the body changes by $\delta\mathbf{R}_0$, and so the change in the potential energy is

$$\delta U = \sum(\partial U/\partial\mathbf{R})\cdot\delta\mathbf{R} = \delta\mathbf{R}_0\cdot\sum\partial U/\partial\mathbf{R} = -\delta\mathbf{R}_0\cdot\sum\mathbf{f} = -\mathbf{F}\cdot\delta\mathbf{R}_0.$$

Let us now derive the second equation of motion, which gives the time derivative of the angular momentum $\mathbf{M}$. To simplify the derivation, it is convenient to choose the "fixed" (inertial) frame of reference in such a way that the centre of mass is at rest in that frame at the instant considered. The equation of motion thus derived is valid in any other inertial frame, by Galileo's relativity principle.

We have $\dot{\mathbf{M}} = (\mathrm{d}/\mathrm{d}t)\Sigma\mathbf{r}\times\mathbf{p} = \Sigma\dot{\mathbf{r}}\times\mathbf{p}+\Sigma\mathbf{r}\times\dot{\mathbf{p}}$. Our choice of the frame of reference (with $\mathbf{V} = 0$) means that the value of $\dot{\mathbf{r}}$ at the instant considered is the same as $\mathbf{v} = \dot{\mathbf{R}}$. Since the vectors $\mathbf{v}$ and $\mathbf{p} = m\mathbf{v}$ are parallel, $\dot{\mathbf{r}}\times\mathbf{p} = 0$. Replacing $\dot{\mathbf{p}}$ by the force $\mathbf{f}$, we have finally

$$\mathrm{d}\mathbf{M}/\mathrm{d}t = \mathbf{K}, \tag{27.3}$$

where

$$\mathbf{K} = \sum \mathbf{r}\times\mathbf{f}. \tag{27.4}$$

The vector $\mathbf{r}\times\mathbf{f}$ is called the *moment* of the force $\mathbf{f}$, and so $\mathbf{K}$ is the total *torque*, i.e. the sum of the moments of all the forces acting on the body. Like the total force $\mathbf{F}$, the sum (27.4) need include only the external forces: by the law of conservation of angular momentum, the sum of the moments of the internal forces in a closed system must be zero.

The moment of a force, like the angular momentum, in general depends on the choice of the origin about which it is defined. In (27.3) and (27.4) the moments are defined with respect to the centre of mass of the body.

When the origin is moved a distance $\mathbf{a}$, the new position vector $\mathbf{r}'$ of each point in the body is equal to $\mathbf{r}-\mathbf{a}$. Hence $\mathbf{K} = \Sigma\mathbf{r}\times\mathbf{f} = \Sigma\mathbf{r}'\times\mathbf{f}+$ $+\Sigma\mathbf{a}\times\mathbf{f}$ or

$$\mathbf{K} = \mathbf{K}'+\mathbf{a}\times\mathbf{F}. \tag{27.5}$$

Hence we see, in particular, that the value of the torque is independent of the choice of origin if the total force $\mathbf{F} = 0$. In this case the body is said to be acted on by a *couple*.

The change in the potential energy resulting from an infinitesimal rotation $\delta\boldsymbol{\phi}$ of the body is $\delta U = -\Sigma\mathbf{f}\cdot\delta\mathbf{R} = -\Sigma\mathbf{f}\cdot\delta\boldsymbol{\phi}\times\mathbf{r} = -\delta\boldsymbol{\phi}\cdot$ $\cdot\Sigma\mathbf{r}\times\mathbf{f} = -\mathbf{K}\cdot\delta\boldsymbol{\phi}$, whence

$$\mathbf{K} = -\partial U/\partial\boldsymbol{\phi}, \tag{27.6}$$

which is analogous to formula (27.2) for the force.

Let us assume that the vectors **F** and **K** are perpendicular. Then a vector **a** can always be found such that **K**′ given by formula (27.5) is zero and

$$\mathbf{K} = \mathbf{a} \times \mathbf{F}. \tag{27.7}$$

The choice of **a** is not unique, since the addition to **a** of any vector parallel to **F** does not affect equation (27.7). The condition $\mathbf{K}' = 0$ thus gives a straight line, not a point, in the moving system of coordinates. When **K** is perpendicular to **F**, the effect of all the applied forces can therefore be reduced to that of a single force **F** acting along this line.

Such a case is that of a uniform field of force, in which the force on a particle is $\mathbf{f} = e\mathbf{E}$, with **E** a constant vector characterising the field and e characterising the properties of a particle with respect to the field.[†] Then $\mathbf{F} = \mathbf{E}\Sigma e$, $\mathbf{K} = \Sigma e\mathbf{r} \times \mathbf{E}$. Assuming that $\Sigma e \neq 0$, we define a position $\mathbf{r}_0$ such that

$$\mathbf{r}_0 = \sum e\mathbf{r} / \sum e. \tag{27.8}$$

Then the total torque is simply

$$\mathbf{K} = \mathbf{r}_0 \times \mathbf{F}. \tag{27.9}$$

Thus, when a rigid body moves in a uniform field, the effect of the field reduces to the action of a single force **F** applied at the point whose position is (27.8). The position of this point is entirely determined by the properties of the body itself. In a gravitational field, for example, it is the centre of mass.

§28. Rigid bodies in contact

The equations of motion (27.1) and (27.3) show that the conditions of equilibrium for a rigid body can be written as the vanishing of the total force and total torque on the body:

$$\mathbf{F} = \sum \mathbf{f} = 0, \quad \mathbf{K} = \sum \mathbf{r} \times \mathbf{f} = 0. \tag{28.1}$$

---

† For example, in a uniform electric field **E** is the field strength and e the charge; in a uniform gravitational field **E** is the acceleration **g** due to gravity and e is the mass m.

Here the summation is over all the external forces acting on the body, and $\mathbf{r}$ is the position of the "point of application"; the origin with respect to which the torque is defined may be chosen arbitrarily, since if $\mathbf{F} = 0$ the value of $\mathbf{K}$ does not depend on this choice.

If we have a system of rigid bodies in contact, the conditions (28.1) for each body separately must hold in equilibrium. The forces considered must include those exerted on each body by those with which it is in contact. These forces at the points of contact are called *reactions*. It is obvious that the mutual reactions of any two bodies are equal in magnitude and opposite in direction.

In general, both the magnitudes and the directions of the reactions are found by solving simultaneously the equations of equilibrium (28.1) for all the bodies. In some cases, however, their directions are given by the conditions of the problem. For example, if two bodies can slide freely on each other, the reaction between them is normal to the surface.

If two bodies in contact are in relative motion, dissipative forces of *friction* arise, in addition to the reaction.

There are two possible types of motion of bodies in contact—*sliding* and *rolling*. In sliding, the reaction is perpendicular to the surfaces in contact, and the friction is tangential. Pure rolling, on the other hand, is characterised by the fact that there is no relative motion of the bodies at the point of contact; that is, a rolling body is at every instant as it were fixed to the point of contact. The reaction may be in any direction, i.e. it need not be normal to the surfaces in contact. The friction in rolling appears as an additional torque which opposes rolling.

If the friction in sliding is negligibly small, the surfaces concerned are said to be *perfectly smooth*. If, on the other hand, only pure rolling without sliding is possible, and the friction in rolling can be neglected, the surfaces are said to be *perfectly rough*.

In both these cases the frictional forces do not appear explicitly in the problem, which is therefore purely one of mechanics. If, on the other hand, the properties of the friction play an essential part in determining the motion, then the latter is not a purely mechanical process (cf. §20).

Contact between two bodies reduces the number of their degrees of freedom as compared with the case of free motion. Hitherto, in discussing such problems, we have taken this reduction into account by using coordinates which correspond directly to the actual number of degrees of freedom. In rolling, however, such a choice of coordinates may be impossible.

The condition imposed on the motion of rolling bodies is that the velocities of the points in contact should be equal; for example, when a body rolls on a fixed surface, the velocity of the point of contact must be zero. In the general case, this condition is expressed by the *equations of constraint*, of the form

$$\sum_i c_{\alpha i}\dot{q}_i = 0, \tag{28.2}$$

where the $c_{\alpha i}$ are functions of the coordinates only, and the suffix α denumerates the equations. If the left-hand sides of these equations are not the total time derivatives of some functions of the coordinates, the equations cannot be integrated. In other words, they cannot be reduced to relations between the coordinates only, which could be used to express the position of the bodies in terms of fewer coordinates, corresponding to the actual number of degrees of freedom. Such constraints are said to be *non-holonomic*, as opposed to *holonomic* constraints, which impose relations between the coordinates only.

Let us consider, for example, the rolling of a sphere on a plane. As usual, we denote by $\mathbf{V}$ the translational velocity (the velocity of the centre of the sphere), and by $\mathbf{\Omega}$ the angular velocity of rotation. The velocity of the point of contact with the plane is found by putting $\mathbf{r} = -a\mathbf{n}$ in the general formula $\mathbf{v} = \mathbf{V}+\mathbf{\Omega}\times\mathbf{r}$; a is the radius of the sphere and $\mathbf{n}$ a unit vector along the normal to the plane. The required condition is that there should be no sliding at the point of contact, i.e.

$$\mathbf{V} - a\mathbf{\Omega}\times\mathbf{n} = 0. \tag{28.3}$$

This cannot be integrated: although the velocity $\mathbf{V}$ is the total time derivative of the position vector of the centre of the sphere, the angular velocity is not in general the total time derivative of any coordinate. The constraint (28.3) is therefore non-holonomic.†

There is a method of deriving the equations of motion for bodies in contact, in which the reactions are introduced explicitly. The essential feature of this method, which is sometimes called *d'Alembert's principle*, is to write for each of the bodies in contact the equations

$$d\mathbf{P}/dt = \sum \mathbf{f}, \qquad d\mathbf{M}/dt = \sum \mathbf{r}\times\mathbf{f}, \tag{28.4}$$

wherein the forces **f** acting on each body include the reactions. The latter are initially unknown and are determined, together with the motion of the body, by solving the equations. This method is equally applicable for both holonomic and non-holonomic constraints.

PROBLEMS

PROBLEM 1. Using d'Alembert's principle, find the equations of motion of a homogeneous sphere rolling on a plane under an external force **F** and torque **K**.

SOLUTION. The constraint equation is (28.3). Denoting the reaction force at the point of contact between the sphere and the plane by **R**, we have equations (28.4) in the form

$$\mu\, d\mathbf{V}/dt = \mathbf{F}+\mathbf{R}, \tag{1}$$

$$I\, d\mathbf{\Omega}/dt = \mathbf{K}-a\mathbf{n}\times\mathbf{R}, \tag{2}$$

where we have used the facts that $\mathbf{P} = \mu\mathbf{V}$ and, for a spherical top, $\mathbf{M} = I\mathbf{\Omega}$. Differentiating the constraint equation (28.3) with respect to time, we have $\dot{\mathbf{V}} = a\dot{\mathbf{\Omega}}\times\mathbf{n}$. Substituting in equation (1) and eliminating $\dot{\mathbf{\Omega}}$ by means of (2), we obtain $(I/a\mu)(\mathbf{F}+\mathbf{R}) = \mathbf{K}\times\mathbf{n}-a\mathbf{R}+a\mathbf{n}(\mathbf{n}\cdot\mathbf{R})$, which relates **R**, **F** and **K**. Writing this equation in components and substituting $I = \frac{2}{5}\mu a^2$ (§25, Problem 2(b)), we have

$$R_x = \frac{5}{7a}K_y - \frac{2}{7}F_x, \qquad R_y = -\frac{5}{7a}K_x - \frac{2}{7}F_y, \qquad R_z = -F_z,$$

where the plane is taken as the xy plane. Finally, substituting these expressions in (1), we obtain the equations of motion involving only the given external force and

† It may be noted that the similar constraint in the rolling of a cylinder *is* holonomic. In that case the axis of rotation has a fixed direction in space, and hence $\Omega = d\phi/dt$ is the total derivative of the angle ϕ of rotation of the cylinder about its axis. The condition (28.3) can therefore be integrated, and gives a relation between the angle ϕ and the coordinate of the centre of mass.

torque:

$$\frac{dV_x}{dt} = \frac{5}{7\mu}\left(F_x + \frac{K_y}{a}\right),$$

$$\frac{dV_y}{dt} = \frac{5}{7\mu}\left(F_y - \frac{K_x}{a}\right).$$

The components Ω_x, Ω_y of the angular velocity are given in terms of V_x, V_y by the constraint equation (28.3); for Ω_z we have the equation $\frac{2}{5}\mu a^2 \, d\Omega_z/dt = K_z$, the z component of equation (2).

PROBLEM 2. A rod of weight P has one end A on a vertical plane and the other end B on a horizontal plane (Fig. 25), and is held in position by two horizontal strings AD and BC, the latter being in the same vertical plane as AB. Determine the reactions of the planes and the tensions in the strings.

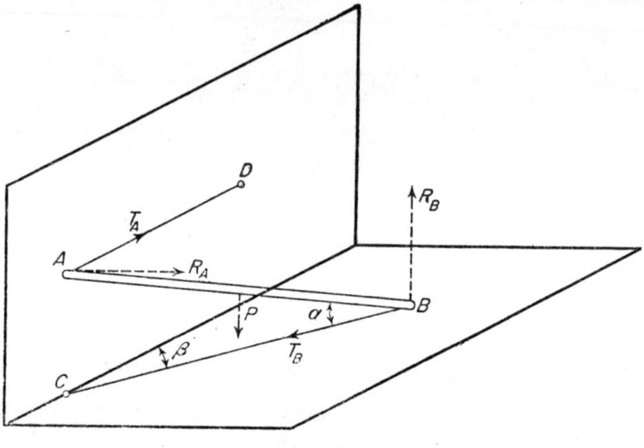

FIG. 25

SOLUTION. The tensions T_A and T_B are from A to D and from B to C respectively. The reactions R_A and R_B are perpendicular to the corresponding planes. The solution of the equations of equilibrium gives $R_B = P$, $T_B = \frac{1}{2}P \cot \alpha$, $R_A = T_B \sin \beta$, $T_A = T_B \cos \beta$.

PROBLEM 3. Two rods of length l and negligible weight are hinged together, and their ends are connected by a string AB (Fig. 26). They stand on a plane, and a force F is applied at the midpoint of one rod. Determine the reactions.

SOLUTION. The tension T acts at A from A to B, and at B from B to A. The reactions R_A and R_B at A and B are perpendicular to the plane. Let $\mathbf{R}_C$ be the reaction on the rod AC at the hinge; then a reaction $-\mathbf{R}_C$ acts on the rod BC. The condition that the sum of the moments of the forces $\mathbf{R}_B$, T and $-\mathbf{R}_C$ acting on

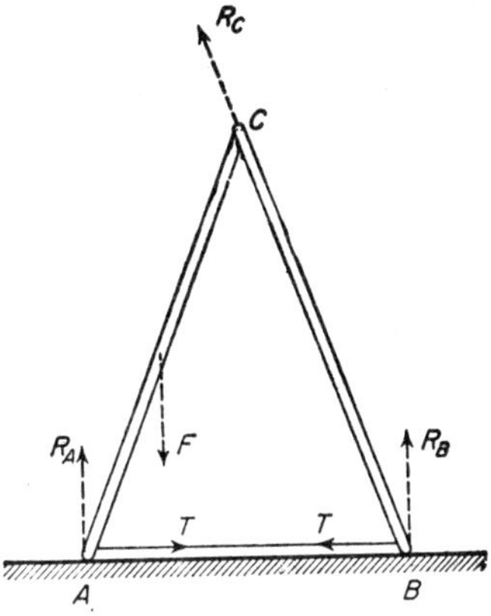

Fig. 26

the rod BC should be zero shows that $\mathbf{R}_C$ acts along BC. The remaining conditions of equilibrium (for the two rods separately) give $R_A = \frac{3}{4}F$, $R_B = \frac{1}{4}F$, $R_C = \frac{1}{4}F$ cosec α, $T = \frac{1}{4}F \cot \alpha$, where α is the angle CAB.

§29. Motion in a non-inertial frame of reference

Up to this point we have always used inertial frames of reference in discussing the motion of mechanical systems. For example, the Lagrangian

$$L_0 = \tfrac{1}{2}m\mathbf{v}_0^2 - U, \qquad (29.1)$$

and the corresponding equation of motion $m\,\mathrm{d}\mathbf{v}_0/\mathrm{d}t = -\partial U/\partial \mathbf{r}$, for a single particle in an external field are valid only in an inertial frame. (In this section the suffix 0 denotes quantities pertaining to an inertial frame.)

Let us now consider what the equations of motion will be in a non-inertial frame of reference. The basis of the solution of this problem is again the principle of least action, whose validity does not depend on the frame of reference chosen. Lagrange's equations

$$\frac{\mathrm{d}}{\mathrm{d}t}\left(\frac{\partial L}{\partial \mathbf{v}}\right) = \frac{\partial L}{\partial \mathbf{r}} \qquad (29.2)$$

are likewise valid, but the Lagrangian is no longer of the form (29.1), and to derive it we must carry out the necessary transformation of the function L_0.

This transformation is done in two steps. Let us first consider a frame of reference K' which moves with a translational velocity $\mathbf{V}(t)$ relative to the inertial frame K_0. The velocities $\mathbf{v}_0$ and $\mathbf{v}'$ of a particle in the frames K_0 and K' respectively are related by

$$\mathbf{v}_0 = \mathbf{v}' + \mathbf{V}(t). \tag{29.3}$$

Substitution of this in (29.1) gives the Lagrangian in K':

$$L' = \tfrac{1}{2}m\mathbf{v}'^2 + m\mathbf{v}'\cdot\mathbf{V} + \tfrac{1}{2}m\mathbf{V}^2 - U.$$

Now $\mathbf{V}^2(t)$ is a given function of time, and can be written as the total derivative with respect to t of some other function; the third term in L' can therefore be omitted. Next, $\mathbf{v}' = \mathrm{d}\mathbf{r}'/\mathrm{d}t$, where $\mathbf{r}'$ is the position of the particle in the frame K'. Hence

$$m\mathbf{V}(t)\cdot\mathbf{v}' = m\mathbf{V}\cdot\mathrm{d}\mathbf{r}'/\mathrm{d}t = \mathrm{d}(m\mathbf{V}\cdot\mathbf{r}')/\mathrm{d}t - m\mathbf{r}'\cdot\mathrm{d}\mathbf{V}/\mathrm{d}t.$$

Substituting in the Lagrangian and again omitting the total time derivative, we have finally

$$L' = \tfrac{1}{2}mv'^2 - m\mathbf{W}(t)\cdot\mathbf{r}' - U, \tag{29.4}$$

where $\mathbf{W} = \mathrm{d}\mathbf{V}/\mathrm{d}t$ is the translational acceleration of the frame K'.

The Lagrange's equation derived from (29.4) is

$$m\frac{\mathrm{d}\mathbf{v}'}{\mathrm{d}t} = -\frac{\partial U}{\partial \mathbf{r}'} - m\mathbf{W}(t). \tag{29.5}$$

Thus an accelerated translational motion of a frame of reference is equivalent, as regards its effect on the equations of motion of a particle, to the application of a uniform field of force equal to the mass of the particle multiplied by the acceleration $\mathbf{W}$, in the direction opposite to this acceleration.

Let us now bring in a further frame of reference K, whose origin coincides with that of K', but which rotates relative to K' with angular velocity $\mathbf{\Omega}(t)$. Thus K executes both a translational and a rotational motion relative to the inertial frame K_0.

The velocity $\mathbf{v}'$ of the particle relative to K' is composed of its velocity $\mathbf{v}$ relative to K and the velocity $\mathbf{\Omega}\times\mathbf{r}$ of its rotation with K: $\mathbf{v}' = \mathbf{v}+\mathbf{\Omega}\times\mathbf{r}$ (since the positions $\mathbf{r}$ and $\mathbf{r}'$ in the frames K and K' coincide). Substituting this in the Lagrangian (29.4), we obtain

$$L = \tfrac{1}{2}mv^2+m\mathbf{v}\cdot\mathbf{\Omega}\times\mathbf{r}+\tfrac{1}{2}m(\mathbf{\Omega}\times\mathbf{r})^2-m\mathbf{W}\cdot\mathbf{r}-U. \tag{29.6}$$

This is the general form of the Lagrangian of a particle in an arbitrary, not necessarily inertial, frame of reference. The rotation of the frame leads to the appearance in the Lagrangian of a term linear in the velocity of the particle.

To calculate the derivatives appearing in Lagrange's equation, we write the total differential

$$\begin{aligned} \mathrm{d}L &= m\mathbf{v}\cdot\mathrm{d}\mathbf{v}+m\mathrm{d}\mathbf{v}\cdot\mathbf{\Omega}\times\mathbf{r}+m\mathbf{v}\cdot\mathbf{\Omega}\times\mathrm{d}\mathbf{r}+ \\ &\quad +m(\mathbf{\Omega}\times\mathbf{r})\cdot(\mathbf{\Omega}\times\mathrm{d}\mathbf{r})-m\mathbf{W}\cdot\mathrm{d}\mathbf{r}-(\partial U/\partial\mathbf{r})\cdot\mathrm{d}\mathbf{r} \\ &= m\mathbf{v}\cdot\mathrm{d}\mathbf{v}+m\,\mathrm{d}\mathbf{v}\cdot\mathbf{\Omega}\times\mathbf{r}+m\,\mathrm{d}\mathbf{r}\cdot\mathbf{v}\times\mathbf{\Omega}+ \\ &\quad +m(\mathbf{\Omega}\times\mathbf{r})\times\mathbf{\Omega}\cdot\mathrm{d}\mathbf{r}-m\mathbf{W}\cdot\mathrm{d}\mathbf{r}-(\partial U/\partial\mathbf{r})\cdot\mathrm{d}\mathbf{r}. \end{aligned}$$

The terms in $\mathrm{d}\mathbf{v}$ and $\mathrm{d}\mathbf{r}$ give

$$\begin{aligned} \partial L/\partial\mathbf{v} &= m\mathbf{v}+m\mathbf{\Omega}\times\mathbf{r}, \\ \partial L/\partial\mathbf{r} &= m\mathbf{v}\times\mathbf{\Omega}+m(\mathbf{\Omega}\times\mathbf{r})\times\mathbf{\Omega}-m\mathbf{W}-\partial U/\partial\mathbf{r}. \end{aligned}$$

Substitution of these expressions in (29.2) gives the required equation of motion:

$$m\,\mathrm{d}\mathbf{v}/\mathrm{d}t = -\partial U/\partial\mathbf{r}-m\mathbf{W}+m\mathbf{r}\times\dot{\mathbf{\Omega}}+2m\mathbf{v}\times\mathbf{\Omega}+m\mathbf{\Omega}\times(\mathbf{r}\times\mathbf{\Omega}). \tag{29.7}$$

We see that the "inertia forces" due to the rotation of the frame consist of three terms. The force $m\mathbf{r}\times\dot{\mathbf{\Omega}}$ is due to the non-uniformity of the rotation, but the other two terms appear even if the rotation is uniform. The force $2m\mathbf{v}\times\mathbf{\Omega}$ is called the *Coriolis force*; unlike any other (non-dissipative) force hitherto considered, it depends on the velocity of the particle. The force $m\mathbf{\Omega}\times(\mathbf{r}\times\mathbf{\Omega})$ is called the *centrifugal force*. It lies in the plane through $\mathbf{r}$ and $\mathbf{\Omega}$, is perpendicular to the axis of rotation (i.e. to $\mathbf{\Omega}$), and is directed away from the axis. The magnitude of this force is $m\varrho\Omega^2$, where ϱ is the distance of the particle from the axis of rotation.

Let us now consider the particular case of a uniformly rotating frame with no translational acceleration. Putting in (29.6) and (29.7) $\mathbf{\Omega}$ = constant, $\mathbf{W} = 0$, we obtain the Lagrangian

$$L = \tfrac{1}{2}mv^2 + m\mathbf{v}\cdot\mathbf{\Omega}\times\mathbf{r} + \tfrac{1}{2}m(\mathbf{\Omega}\times\mathbf{r})^2 - U \tag{29.8}$$

and the equation of motion

$$m\,\mathrm{d}\mathbf{v}/\mathrm{d}t = -\partial U/\partial\mathbf{r} + 2m\mathbf{v}\times\mathbf{\Omega} + m\mathbf{\Omega}\times(\mathbf{r}\times\mathbf{\Omega}). \tag{29.9}$$

The energy of the particle in this case is obtained by substituting

$$\mathbf{p} = \partial L/\partial\mathbf{v} = m\mathbf{v} + m\mathbf{\Omega}\times\mathbf{r} \tag{29.10}$$

in $E = \mathbf{p}\cdot\mathbf{v} - L$, which gives

$$E = \tfrac{1}{2}mv^2 - \tfrac{1}{2}m(\mathbf{\Omega}\times\mathbf{r})^2 + U. \tag{29.11}$$

It should be noticed that the energy contains no term linear in the velocity. The rotation of the frame simply adds to the energy a term depending only on the coordinates of the particle and proportional to the square of the angular velocity. This additional term $-\tfrac{1}{2}m(\mathbf{\Omega}\times\mathbf{r})^2$ is called the *centrifugal potential energy*.

The velocity $\mathbf{v}$ of the particle relative to the uniformly rotating frame of reference is related to its velocity $\mathbf{v}_0$ relative to the inertial frame K_0 by

$$\mathbf{v}_0 = \mathbf{v} + \mathbf{\Omega}\times\mathbf{r}. \tag{29.12}$$

The momentum $\mathbf{p}$ (29.10) of the particle in the frame K is therefore the same as its momentum $\mathbf{p}_0 = m\mathbf{v}_0$ in the frame K_0. The angular momenta $\mathbf{M}_0 = \mathbf{r}\times\mathbf{p}_0$ and $\mathbf{M} = \mathbf{r}\times\mathbf{p}$ are likewise equal. The energies of the particle in the two frames are not the same, however. Substituting $\mathbf{v}$ from (29.12) in (29.11), we obtain $E = \tfrac{1}{2}mv_0^2 - m\mathbf{v}_0\cdot\mathbf{\Omega}\times\mathbf{r} + U = \tfrac{1}{2}mv_0^2 + U - m\mathbf{r}\times\mathbf{v}_0\cdot\mathbf{\Omega}$. The first two terms are the energy E_0 in the frame K_0. Using the angular momentum **M**, we have

$$E = E_0 - \mathbf{M}\cdot\mathbf{\Omega}. \tag{29.13}$$

This formula gives the law of transformation of energy when we change to a uniformly rotating frame. Although it has been derived for a single particle, the derivation can evidently be generalised immediately to any system of particles, and the same formula (29.13) is obtained.

PROBLEMS

PROBLEM 1. Find the deflection of a freely falling body from the vertical caused by the Earth's rotation, assuming the angular velocity of this rotation to be small.

SOLUTION. In a gravitational field $U = -m\mathbf{g}\cdot\mathbf{r}$, where **g** is the gravity acceleration vector; neglecting the centrifugal force in equation (29.9) as containing the square of $\mathbf{\Omega}$, we have the equation of motion

$$\dot{\mathbf{v}} = 2\mathbf{v}\times\mathbf{\Omega}+\mathbf{g}. \tag{1}$$

This equation may be solved by successive approximations. To do so, we put $\mathbf{v} = \mathbf{v}_1+\mathbf{v}_2$, where $\mathbf{v}_1$ is the solution of the equation $\dot{\mathbf{v}}_1 = \mathbf{g}$, i.e. $\mathbf{v}_1 = \mathbf{g}t+\mathbf{v}_0$ ($\mathbf{v}_0$ being the initial velocity). Substituting $\mathbf{v} = \mathbf{v}_1+\mathbf{v}_2$ in (1) and retaining only $\mathbf{v}_1$ on the right, we have for $\mathbf{v}_2$ the equation $\dot{\mathbf{v}}_2 = 2\mathbf{v}_1\times\mathbf{\Omega} = 2t\mathbf{g}\times\mathbf{\Omega}+2\mathbf{v}_0\times\mathbf{\Omega}$. Integration gives

$$\mathbf{r} = \mathbf{h}+\mathbf{v}_0 t+\tfrac{1}{2}\mathbf{g}t^2+\tfrac{1}{3}t^3\mathbf{g}\times\mathbf{\Omega}+t^2\mathbf{v}_0\times\mathbf{\Omega}, \tag{2}$$

where **h** is the initial position of the particle.

Let the z axis be vertically upwards, and the x axis towards the pole; then $g_x = g_y = 0$, $g_z = -g$; $\Omega_x = \Omega\cos\lambda$, $\Omega_y = 0$, $\Omega_z = \Omega\sin\lambda$, where λ is the latitude (which for definiteness we take to be north). Putting $\mathbf{v}_0 = 0$ in (2), we find $x = 0$, $y = -\frac{1}{3}t^3 g\Omega\cos\lambda$. Substitution of the time of fall $t \approx \sqrt{(2h/g)}$ gives finally $x = 0$, $y = -\frac{1}{3}(2h/g)^{3/2}g\Omega\cos\lambda$, the negative value indicating an eastward deflection.

PROBLEM 2. Determine the deflection from coplanarity of the path of a particle thrown from the Earth's surface with velocity $\mathbf{v}_0$.

SOLUTION. Let the xz plane be such as to contain the velocity $\mathbf{v}_0$. The initial altitude $h = 0$. The lateral deviation is given by (2), Problem 1: $y = -\frac{1}{3}t^3 g\Omega_x+t^2(\Omega_x v_{0z}-\Omega_z v_{0x})$ or, substituting the time of flight $t \approx 2v_{0z}/g$, $y = 4v_{0z}^2\Omega(\frac{1}{3}v_{0z}\cos\lambda - v_{0x}\sin\lambda)/g^2$.

PROBLEM 3. Determine the effect of the Earth's rotation on small oscillations of a pendulum (the problem of *Foucault's pendulum*).

SOLUTION. Neglecting the vertical displacement of the pendulum, as being a quantity of the second order of smallness, we can regard the motion as taking place in the horizontal xy plane. Omitting terms in Ω^2, we have the equations of motion $\ddot{x}+\omega^2 x = 2\Omega_z \dot{y}$, $\ddot{y}+\omega^2 y = -2\Omega_z \dot{x}$, where ω is the frequency of oscillation of the pendulum if the Earth's rotation is neglected. Multiplying the second equation by i and adding, we obtain a single equation $\ddot{\xi}+2i\Omega_z\dot{\xi}+\omega^2\xi = 0$ for the complex quantity $\xi = x+iy$. For $\Omega_z \ll \omega$, the solution of this equation is

$$\xi = \exp(-i\Omega_z t)\,[A_1 \exp(i\omega t)+A_2 \exp(-i\omega t)]$$

or

$$x+iy = (x_0+iy_0)\exp(-i\Omega_z t),$$

where the functions $x_0(t)$, $y_0(t)$ give the path of the pendulum when the Earth's rotation is neglected. The effect of this rotation is therefore to turn the path about the vertical with angular velocity Ω_z.

CHAPTER 7

THE CANONICAL EQUATIONS†

§30. Hamilton's equations

The formulation of the laws of mechanics in terms of the Lagrangian, and of Lagrange's equations derived from it, presupposes that the mechanical state of a system is described by specifying its generalised coordinates and velocities. This is not the only possible mode of description, however. A number of advantages, especially in the study of certain general problems of mechanics, attach to a description in terms of the generalised coordinates and momenta of the system. The question therefore arises of the form of the equations of motion corresponding to that formulation of mechanics.

The passage from one set of independent variables to another can be effected by means of what is called in mathematics *Legendre's transformation*. In the present case this transformation is as follows. The total differential of the Lagrangian as a function of coordinates and

† The reader may find useful the following table showing certain differences between the nomenclature used in this book and that which is generally used in the English literature.

Here	*Elsewhere*
Principle of least action	Hamilton's principle
Action	Hamilton's principal function
Abbreviated action	Action

—*Translators.*

velocities is

$$dL = \sum_i \frac{\partial L}{\partial q_i} dq_i + \sum_i \frac{\partial L}{\partial \dot{q}_i} d\dot{q}_i .$$

This expression may be written

$$dL = \sum \dot{p}_i \, dq_i + \sum p_i \, d\dot{q}_i , \tag{30.1}$$

since the derivatives $\partial L/\partial \dot{q}_i$ are, by definition, the generalised momenta, and $\partial L/\partial q_i = \dot{p}_i$ by Lagrange's equations. Writing the second term in (30.1) as $\Sigma p_i \, d\dot{q}_i = d(\Sigma p_i \dot{q}_i) - \Sigma \dot{q}_i \, dp_i$, taking the differential $d(\Sigma p_i \dot{q}_i)$ to the left-hand side, and reversing the signs, we obtain from (30.1)

$$d(\sum p_i \dot{q}_i - L) = -\sum \dot{p}_i \, dq_i + \sum \dot{q}_i \, dp_i .$$

The argument of the differential is the energy of the system (cf. §6); expressed in terms of coordinates and momenta, it is called the *Hamilton's function* or *Hamiltonian* of the system:

$$H(p, q, t) = \sum_i p_i \dot{q}_i - L. \tag{30.2}$$

From the equation in differentials

$$dH = -\sum \dot{p}_i \, dq_i + \sum \dot{q}_i \, dp_i , \tag{30.3}$$

we have the equations

$$\dot{q}_i = \partial H/\partial p_i, \quad \dot{p}_i = -\partial H/\partial q_i. \tag{30.4}$$

These are the required equations of motion in the variables p and q, and are called *Hamilton's equations.* They form a set of $2s$ first-order differential equations for the $2s$ unknown functions $p_i(t)$ and $q_i(t)$, replacing the s second-order equations in the Lagrangian treatment. Because of their simplicity and symmetry of form, they are also called *canonical equations.*

The total time derivative of the Hamiltonian is

$$\frac{dH}{dt} = \frac{\partial H}{\partial t} + \sum \frac{\partial H}{\partial q_i} \dot{q}_i + \sum \frac{\partial H}{\partial p_i} \dot{p}_i .$$

Substitution of $\dot{q}_i$ and $\dot{p}_i$ from equations (30.4) shows that the last two terms cancel, and so

$$\mathrm{d}H/\mathrm{d}t = \partial H/\partial t. \tag{30.5}$$

In particular, if the Hamiltonian does not depend explicitly on time, then $\mathrm{d}H/\mathrm{d}t = 0$, and we have the law of conservation of energy.

As well as the dynamical variables q, $\dot{q}$ or q, p, the Lagrangian and the Hamiltonian involve various parameters which relate to the properties of the mechanical system itself, or to the external forces on it. Let λ be one such parameter. Regarding it as a variable, we have instead of (30.1)

$$\mathrm{d}L = \sum \dot{p}_i\,\mathrm{d}q_i + \sum p_i\,\mathrm{d}\dot{q}_i + (\partial L/\partial\lambda)\,\mathrm{d}\lambda,$$

and (30.3) becomes

$$\mathrm{d}H = -\sum \dot{p}_i\,\mathrm{d}q_i + \sum \dot{q}_i\,\mathrm{d}p_i - (\partial L/\partial\lambda)\,\mathrm{d}\lambda.$$

Hence

$$(\partial H/\partial\lambda)_{p,q} = -(\partial L/\partial\lambda)_{\dot{q},q}, \tag{30.6}$$

which relates the derivatives of the Lagrangian and the Hamiltonian with respect to the parameter λ. The suffixes to the derivatives show the quantities which are to be kept constant in the differentiation.

This result can be put in another way. Let the Lagrangian be of the form $L = L_0 + L'$, where L' is a small correction to the function L_0. Then the corresponding addition H' in the Hamiltonian $H = H_0 + H'$ is related to L' by

$$(H')_{p,q} = -(L')_{\dot{q},q}. \tag{30.7}$$

PROBLEMS

Problem 1. Find the Hamiltonian for a single particle in Cartesian, cylindrical and spherical coordinates.

Solution. In Cartesian coordinates x, y, z,

$$H = \frac{1}{2m}(p_x^2 + p_y^2 + p_z^2) + U(x, y, z);$$

in cylindrical coordinates r, ϕ, z,

$$H = \frac{1}{2m}\left(p_r^2 + \frac{p_\phi^2}{r^2} + p_z^2\right) + U(r, \phi, z);$$

in spherical coordinates r, θ, ϕ,

$$H = \frac{1}{2m}\left(p_r^2 + \frac{p_\theta^2}{r^2} + \frac{p_\phi^2}{r^2 \sin^2\theta}\right) + U(r, \theta, \phi).$$

PROBLEM 2. Find the Hamiltonian for a particle in a uniformly rotating frame of reference.

SOLUTION. Expressing the velocity $\mathbf{v}$ in the energy (29.11) in terms of the momentum $\mathbf{p}$ by (29.10), we have $H = p^2/2m - \mathbf{\Omega}\cdot\mathbf{r}\times\mathbf{p} + U$.

§31. The Hamilton–Jacobi equation

In formulating the principle of least action, we have considered the integral

$$S = \int_{t_1}^{t_2} L\,\mathrm{d}t, \tag{31.1}$$

taken along a path between two given positions $q^{(1)}$ and $q^{(2)}$ which the system occupies at given instants t_1 and t_2. In varying the action, we compared the values of this integral for neighbouring paths with the same values of $q(t_1)$ and $q(t_2)$. Only one of these paths corresponds to the actual motion, namely the path for which the integral S has its minimum value.

Let us now consider another aspect of the concept of action, regarding S as a quantity characterising the motion along the actual path, and compare the values of S for paths having a common beginning at $q(t_1) = q^{(1)}$, but passing through different points at time t_2. In other words, we consider the action integral for the true path as a function of the coordinates at the upper limit of integration.

The change in the action from one path to a neighbouring path is given (if there is one degree of freedom) by the expression (2.5):

$$\delta S = \left[\frac{\partial L}{\partial \dot{q}}\,\delta q\right]_{t_1}^{t_2} + \int_{t_1}^{t_2} \left(\frac{\partial L}{\partial q} - \frac{\mathrm{d}}{\mathrm{d}t}\frac{\partial L}{\partial \dot{q}}\right)\delta q\,\mathrm{d}t.$$

Since the paths of actual motion satisfy Lagrange's equations, the integral in δS is zero. In the first term we put $\delta q(t_1) = 0$, and denote the value of $\delta q(t_2)$ by δq simply. Replacing $\partial L/\partial \dot{q}$ by p, we have finally $\delta S = p\delta q$ or, in the general case of any number of degrees of freedom,

$$\delta S = \sum_i p_i \delta q_i . \tag{31.2}$$

From this relation it follows that the partial derivatives of the action with respect to the coordinates are equal to the corresponding momenta:

$$\partial S/\partial q_i = p_i . \tag{31.3}$$

The action may similarly be regarded as an explicit function of time, by considering paths starting at a given instant t_1 and at a given point $q^{(1)}$, and ending at a given point $q^{(2)}$ at various times $t_2 = t$. The partial derivative $\partial S/\partial t$ thus obtained may be found by an appropriate variation of the integral. It is simpler, however, to use formula (31.3), proceeding as follows.

From the definition of the action, its total time derivative along the path is

$$\mathrm{d}S/\mathrm{d}t = L. \tag{31.4}$$

Next, regarding S as a function of coordinates and time, in the sense described above, and using formula (31.3), we have

$$\frac{\mathrm{d}S}{\mathrm{d}t} = \frac{\partial S}{\partial t} + \sum \frac{\partial S}{\partial q_i} \dot{q}_i = \frac{\partial S}{\partial t} + \sum_i p_i \dot{q}_i .$$

A comparison gives $\partial S/\partial t = L - \Sigma p_i \dot{q}_i$ or

$$\partial S/\partial t = -H(p, q, t). \tag{31.5}$$

Formulae (31.3) and (31.5) may be represented by the expression

$$\mathrm{d}S = \sum_i p_i \, \mathrm{d}q_i - H \, \mathrm{d}t \tag{31.6}$$

for the total differential of the action as a function of coordinates and time at the upper limit of integration in (31.1).

The action itself correspondingly becomes

$$S = \int \Big(\sum_i p_i \, \mathrm{d}q_i - H \, \mathrm{d}t\Big). \tag{31.7}$$

In particular, if the function $H(p, q)$ does not depend explicitly on the time, so that the energy is conserved, we can replace $H(p, q)$ by a constant E, and the time dependence of S reduces to a term $-Et$:

$$S(q, t) = S_0(q) - Et, \tag{31.8}$$

where

$$S_0(q) = \int \sum_i p_i \, \mathrm{d}q_i \tag{31.9}$$

is sometimes called the *abbreviated action*.

Replacing the momenta p in (31.5) by the derivatives $\partial S/\partial q$, we have the differential equation

$$\frac{\partial S}{\partial t} + H\left(\frac{\partial S}{\partial q_1}, \ldots, \frac{\partial S}{\partial q_s}; q_1, \ldots, q_s; t\right) = 0 \tag{31.10}$$

which must be satisfied by the function $S(q, t)$. This first-order partial differential equation is called the *Hamilton–Jacobi equation*. For a single particle in an external field $U(x, y, z, t)$, for instance, it is

$$\frac{\partial S}{\partial t} + \frac{1}{2m}\left[\left(\frac{\partial S}{\partial x}\right)^2 + \left(\frac{\partial S}{\partial y}\right)^2 + \left(\frac{\partial S}{\partial z}\right)^2\right] + U(x, y, z, t) = 0. \tag{31.11}$$

The Hamilton–Jacobi equation takes a somewhat simpler form if the function $H(p, q)$ does not involve the time explicitly. Taking $S(q, t)$ from (31.8) gives for the abbreviated action $S_0(q)$ the equation

$$H\left(\frac{\partial S_0}{\partial q_1}, \ldots, \frac{\partial S_0}{\partial q_s}; q_1, \ldots, q_s\right) = E. \tag{31.12}$$

§32. Adiabatic invariants

Let us consider a mechanical system executing a finite motion in one dimension and characterised by some parameter λ which specifies the properties of the system or of the external field in which it is placed, and let us suppose that λ varies slowly (*adiabatically*) with time as the result of some external action; by a "slow" variation we mean one in which λ varies only slightly during the period T of the motion:

$$T\,\mathrm{d}\lambda/\mathrm{d}t \ll \lambda. \tag{32.1}$$

Such a system is not closed, and its energy E is not conserved. However, since λ varies only slowly, the rate of change $\dot{E}$ of the energy is proportional to the rate of change $\dot{\lambda}$ of the parameter. This means that the energy of the system behaves as some function of λ when the latter varies. In other words, there is some combination of E and λ which remains constant during the motion. This quantity is called an *adiabatic invariant*.

Let $H(p, q; \lambda)$ be the Hamiltonian of the system, which depends on the parameter λ. According to formula (30.5), the total time derivative of the energy of the system is $\mathrm{d}E/\mathrm{d}t = \partial H/\partial t = (\partial H/\partial\lambda)(\mathrm{d}\lambda/\mathrm{d}t)$. In averaging this equation over the period of the motion, we need not average the second factor, since λ (and therefore $\dot{\lambda}$) varies only slowly: $\overline{\mathrm{d}E/\mathrm{d}t} = (\mathrm{d}\lambda/\mathrm{d}t)\overline{(\partial H/\partial\lambda)}$, and in the averaged function $\overline{\partial H/\partial\lambda}$ we can regard only p and q, and not λ, as variable. That is, the averaging is taken over the motion which would occur if λ remained constant.

The averaging may be explicitly written

$$\overline{\frac{\mathrm{d}E}{\mathrm{d}t}} = \frac{\mathrm{d}\lambda}{\mathrm{d}t}\cdot\frac{1}{T}\int_0^T \frac{\partial H}{\partial\lambda}\,\mathrm{d}t.$$

According to Hamilton's equation $\dot{q} = \partial H/\partial p$, or $\mathrm{d}t = \mathrm{d}q/(\partial H/\partial p)$. The integration with respect to time can therefore be replaced by one with respect to the coordinate, with the period T written as

$$T = \int_0^T \mathrm{d}t = \oint \mathrm{d}q/(\partial H/\partial p);$$

here the $\oint$ sign denotes an integration over the complete range of variation ("there and back") of the coordinate during the period.

Thus

$$\frac{\overline{\mathrm{d}E}}{\mathrm{d}t} = \frac{\mathrm{d}\lambda}{\mathrm{d}t}\frac{\oint (\partial H/\partial\lambda)\,\mathrm{d}q/(\partial H/\partial p)}{\oint \mathrm{d}q/(\partial H/\partial p)}. \tag{32.2}$$

As has already been mentioned, the integrations in this formula must be taken over the path for a given constant value of λ. Along such a path the Hamiltonian has a constant value E, and the momentum is a definite function of the variable coordinate q and of the two independent constant parameters E and λ. Putting therefore $p = p(q; E, \lambda)$ and differentiating with respect to λ the equation $H(p, q; \lambda) = E$, we have $\partial H/\partial\lambda + (\partial H/\partial p)(\partial p/\partial\lambda) = 0$, or

$$\frac{\partial H/\partial\lambda}{\partial H/\partial p} = -\frac{\partial p}{\partial\lambda}.$$

Substituting this in the numerator of (32.2) and writing the integrand in the denominator as $\partial p/\partial E$, we obtain

$$\frac{\overline{\mathrm{d}E}}{\mathrm{d}t} = -\frac{\mathrm{d}\lambda}{\mathrm{d}t}\frac{\oint (\partial p/\partial\lambda)\,\mathrm{d}q}{\oint (\partial p/\partial E)\,\mathrm{d}q} \tag{32.3}$$

or

$$\oint \left(\frac{\partial p}{\partial E}\frac{\overline{\mathrm{d}E}}{\mathrm{d}t} + \frac{\partial p}{\partial\lambda}\frac{\mathrm{d}\lambda}{\mathrm{d}t}\right)\mathrm{d}q = 0.$$

Finally, this may be written as

$$\overline{\mathrm{d}I/\mathrm{d}t} = 0, \tag{32.4}$$

where

$$I \equiv \oint p\,\mathrm{d}q/2\pi, \tag{32.5}$$

the integral being taken over the path for given E and λ. This shows that, in the approximation here considered, I remains constant when the parameter λ varies, i.e. I is an adiabatic invariant.

The integral (32.5) has a geometrical significance in terms of the *phase path* of the system, i.e. the curve which represents the dependence of p on q; the phase path of a system executing a periodic motion is a closed curve. The integral (32.5) taken round this curve is the area enclosed.

As an example, let us determine the adiabatic invariant for a one-dimensional oscillator. The Hamiltonian is $H = \frac{1}{2}p^2/m + \frac{1}{2}m\omega^2 q^2$, where ω is the frequency of the oscillator. The equation of the phase path is given by the law of conservation of energy $H(p, q) = E$. The path is an ellipse with semiaxes $\sqrt{(2mE)}$ and $\sqrt{(2E/m\omega^2)}$, and its area, divided by 2π, is

$$I = E/\omega. \tag{32.6}$$

The adiabatic invariance of I signifies that, when the parameters of the oscillator vary slowly, the energy is proportional to the frequency.

CHAPTER 8

THE PRINCIPLE OF RELATIVITY

§33. Velocity of propagation of interactions

The interaction of material particles is described in ordinary mechanics by means of a potential energy of interaction, which is a function of the coordinates of the interacting particles. It is easy to see that this manner of describing interactions contains the assumption of instantaneous propagation of interactions. For the forces exerted on each of the particles by the other particles at a particular instant of time depend, according to this description, only on the positions of the particles at this one instant. A change in the position of any of the interacting particles influences the other particles immediately.

However, experiment shows that instantaneous interactions do not exist in nature. Thus a mechanics based on the assumption of instantaneous propagation of interactions contains within itself a certain inaccuracy. In actuality, if any change takes place in one of the interacting bodies, it will influence the other bodies only after the lapse of a certain interval of time. Dividing the distance between the two bodies by this time interval, we obtain the *velocity of propagation of the interaction.*

This velocity should, strictly speaking, be called the *maximum* velocity of propagation of interactions. It determines only that interval of time after which a change occurring in one body begins to manifest itself in another by the arrival of a *signal*. It is clear that the existence of a maximum velocity of propagation of interactions implies, at the same time, that motions of bodies with greater velocity than this are impossible in nature.

From the principle of relativity it follows that the velocity of propagation of interactions is the *same* in *all* inertial frames of reference. Thus the velocity of propagation of interactions is a universal constant.

This constant velocity (as we shall show later) is also the velocity of light in empty space. The velocity of light is usually designated by the letter c, and its numerical value is

$$c = 2{\cdot}998 \times 10^{10} \text{ cm/sec.} \tag{33.1}$$

The large value of this velocity explains the fact that in practice classical mechanics is found to be sufficiently accurate in most cases. The velocities with which we have occasion to deal are usually so small compared with the velocity of light that the assumption that the latter is infinite does not materially affect the accuracy of the results.

The combination of the principle of relativity with the finiteness of the velocity of propagation of interactions is called *Einstein's relativity principle* (it was formulated by Einstein in 1905) in contrast to Galileo's relativity principle, which was based on an infinite velocity of propagation of interactions.

The mechanics based on Einstein's relativity principle (we shall usually refer to it simply as the principle of relativity) is called *relativistic*. In the limiting case when the velocities of the moving bodies are small compared with the velocity of light we can neglect the effect on their motion of the finiteness of the velocity of propagation. Then relativistic mechanics goes over into classical mechanics, based on the assumption of instantaneous propagation of interactions. The limiting transition from relativistic to classical mechanics can be produced formally by the transition to the limit $c \to \infty$ in the formulae of relativistic mechanics.

In classical mechanics distance is already relative, i.e. the spatial relations between different events depend on the frame of reference in which they are described. The statement that two non-simultaneous events occur at one and the same point in space or, in general, at a definite distance from each other, acquires a meaning only when we indicate the frame of reference which is used.

On the other hand, time is absolute in classical mechanics; in other words, the properties of time are assumed to be independent of the frame of reference; there is one time for all frames. This means that if any two phenomena occur simultaneously for any one observer, then they occur simultaneously also for all others. In general, the interval of time between two given events must be identical for all frames of reference.

It is easy to show, however, that the idea of an absolute time is in complete contradiction to Einstein's relativity principle. For this it is sufficient to recall that in classical mechanics, based on the concept of an absolute time, the familiar law of composition of velocities is valid, according to which the velocity of a composite motion is simply equal to the (vector) sum of the velocities of the constituent motions. This law, being universal, should also be applicable to the propagation of interactions. From this it would follow that the velocity of propagation must be different in different inertial frames of reference, in contradiction to the principle of relativity. In this matter experiment completely confirms the principle of relativity. Measurements first performed by Michelson (1881) showed complete lack of dependence of the velocity of light on its direction of propagation, whereas according to classical mechanics the velocity of light should not be the same in the direction of the Earth's motion as it is in the opposite direction.

Thus the principle of relativity leads to the result that time is not absolute. Time elapses differently in different frames of reference. Consequently the statement that a definite time interval has elapsed between two given events acquires meaning only when the reference frame to which this statement applies is indicated. In particular, events which are simultaneous in one reference frame will not be simultaneous in other frames.

To clarify this, it is instructive to consider the following simple example. Let us look at two inertial reference frames K and K' with coordinate axes XYZ and $X'Y'Z'$ respectively, where the frame K' moves relative to K along the axis $X(X')$ (Fig. 27).

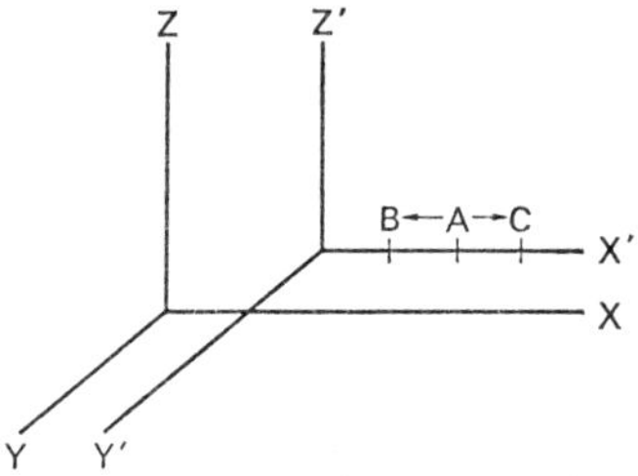

FIG. 27

Suppose signals start out from some point A on the axis X' in two opposite directions. Since the velocity of propagation of a signal in the frame K', as in all inertial frames, is equal (for both directions) to c, the signals will reach points B and C, equidistant from A, at one and the same time (in the frame K'). But it is easy to see that the same two events (arrival of the signal at B and C) can by no means be simultaneous for an observer in the frame K: the velocity of signals relative to the frame K has, according to the principle of relativity, the same value c, and since the point B moves (relative to the frame K) toward the source of its signal, while the point C moves in the direction away from the signal (sent from A to C), in the frame K the signal will reach the point B earlier than the point C.

Thus Einstein's relativity principle introduces fundamental changes in basic physical concepts. The notions of space and time derived by us from our daily experiences are only approximations due to the fact that in daily life we happen to deal only with velocities which are very small compared with the velocity of light.

§34. Intervals

In what follows we shall frequently use the concept of an *event*. An event is described by the place where it occurred and the time when it occurred. Thus an event occurring in a certain material particle is defined by the three coordinates of that particle and the time when the event occurs.

It is frequently useful for reasons of presentation to use a fictitious four-dimensional space, on the axes of which are marked three space coordinates and the time. In this space events are represented by points, called *world points*. In this four-dimensional space there corresponds to each particle a certain line, called a *world line*. The points of this line determine the coordinates of the particle at all moments of time. It is easy to show that to a particle in uniform rectilinear motion there corresponds a straight world line.

We now express the principle of the invariance of the velocity of light in mathematical form. For this purpose we consider two frames of reference K and K' moving relative to each other with constant velocity. We choose the coordinate axes so that the axes X and X' coincide, while the axes Y and Z are parallel to Y' and Z'; we designate the time in the frames K and K' by t and t'.

Let the first event consist of sending out a signal, propagating with light velocity, from a point having coordinates $x_1\, y_1\, z_1$ in the frame K, at time t_1 in this frame. We observe the propagation of this signal in the frame K. Let the second event consist of the arrival of the signal at point $x_2 y_2 z_2$ at the moment of time t_2. The signal propagates with velocity c; the distance covered by it is therefore $c(t_2-t_1)$. On the other hand, this same distance equals $[(x_2-x_1)^2+(y_2-y_1)^2+(z_2-z_1)^2]^{\frac{1}{2}}$. Thus we can write the following relation between the coordinates of the two events in the frame K:

$$(x_2-x_1)^2+(y_2-y_1)^2+(z_2-z_1)^2-c^2(t_2-t_1)^2 = 0. \qquad (34.1)$$

The same two events, i.e. the propagation of the signal, can be observed from the frame K'. Let the coordinates of the first event in the frame K' be $x_1' y_1' z_1' t_1'$, and of the second: $x_2' y_2' z_2' t_2'$. Since the velocity of light is the same in the frames K and K', we have, similarly to (34.1),

$$(x_2'-x_1')^2+(y_2'-y_1')^2+(z_2'-z_1')^2-c^2(t_2'-t_1')^2 = 0. \qquad (34.2)$$

If $x_1 y_1 z_1 t_1$ and $x_2 y_2 z_2 t_2$ are the coordinates of *any* two events, then the quantity

$$s_{12} = [c^2(t_2-t_1)^2-(x_2-x_1)^2-(y_2-y_1)^2-(z_2-z_1)^2]^{\frac{1}{2}} \qquad (34.3)$$

is called the ***interval*** between these two events.

Thus it follows from the principle of invariance of the velocity of light that, if the interval between two events is zero in one frame of reference, then it is equal to zero in all other frames.

If two events are infinitely close to each other, then for the interval $\mathrm{d}s$ between them

$$\mathrm{d}s^2 = c^2\,\mathrm{d}t^2 - \mathrm{d}x^2 - \mathrm{d}y^2 - \mathrm{d}z^2. \tag{34.4}$$

The form of (34.3) and (34.4) permits us to regard the interval, from the formal mathematical point of view, as the distance between two points in a fictitious four-dimensional space (whose coordinates are x, y, z, and the product ct). But there is a basic difference between the rule for forming this quantity and the rule in ordinary geometry: in forming the square of the interval, the squares of the coordinate differences along the different axes are taken, not with the same sign, but rather with varying signs.†

As already shown, if $\mathrm{d}s = 0$ in one inertial frame, then $\mathrm{d}s' = 0$ in any other frame. On the other hand, $\mathrm{d}s$ and $\mathrm{d}s'$ are infinitesimals of the same order. From these two conditions it follows that $\mathrm{d}s^2$ and $\mathrm{d}s'^2$ must be proportional to each other:

$$\mathrm{d}s^2 = a\,\mathrm{d}s'^2,$$

where the coefficient a can depend only on the absolute value of the relative velocity of the two inertial frames. It cannot depend on the coordinates or the time, since then different points in space and different moments in time would not be equivalent, which would be in contradiction to the homogeneity of space and time. Similarly, it cannot depend on the direction of the relative velocity, since that would contradict the isotropy of space.

Let us consider three frames of reference K, K_1, K_2, and let V_1 and V_2 be the velocities of K_1 and K_2 relative to K. We then have

$$\mathrm{d}s^2 = a(V_1)\,\mathrm{d}s_1^2, \quad \mathrm{d}s^2 = a(V_2)\,\mathrm{d}s_2^2.$$

† The four-dimensional geometry described by the quadratic form (34.4) was introduced by Minkowski, in connection with the theory of relativity. This geometry is called *pseudo-Euclidean*, in contrast to ordinary Euclidean geometry.

Similarly we can write

$$ds_1^2 = a(V_{12})\, ds_2^2,$$

where V_{12} is the absolute value of the velocity of K_2 relative to K_1. Comparing these relations with one another, we find that we must have

$$\frac{a(V_2)}{a(V_1)} = a(V_{12}). \tag{34.5}$$

But V_{12} depends not only on the absolute values of the vectors $\mathbf{V}_1$ and $\mathbf{V}_2$, but also on the angle between them. However, this angle does not appear on the left side of formula (34.5). It is therefore clear that this formula can be correct only if the function $a(V)$ reduces to a constant, which is equal to unity according to this same formula.

Thus,

$$ds^2 = ds'^2,$$

and from the equality of the infinitesimal intervals there follows the equality of finite intervals: $s = s'$.

Thus we arrive at a very important result: the interval between two events is the same in all inertial frames of reference, i.e. it is invariant under transformation from one inertial frame to any other. This invariance is the mathematical expression of the constancy of the velocity of light.

Again let $x_1y_1z_1t_1$ and $x_2y_2z_2t_2$ be the coordinates of two events in a certain frame of reference K. Does there exist a frame K' in which these two events occur at one and the same point in space?

We introduce the notation

$$t_2-t_1 = t_{12}, \quad (x_2-x_1)^2+(y_2-y_1)^2+(z_2-z_1)^2 = l_{12}^2.$$

Then the interval between the events in the frame K is

$$s_{12}^2 = c^2t_{12}^2-l_{12}^2$$

and in the frame K'

$$s_{12}'^2 = c^2t_{12}'^2-l_{12}'^2,$$

whereupon, because of the invariance of intervals,

$$c^2t_{12}^2-l_{12}^2 = c^2t_{12}'^2-l_{12}'^2.$$

We want the two events to occur at the same point in the frame K', that is, we require $l'_{12} = 0$. Then

$$s_{12}^2 = c^2t_{12}^2 - l_{12}^2 = c^2t_{12}'^2 > 0.$$

Consequently a frame of reference with the required property exists if $s_{12}^2 > 0$, that is, if the interval between the two events is a real number. Real intervals are said to be *time-like*.

Thus, if the interval between two events is time-like, then there exists a frame of reference in which the two events occur at one and the same place. The time which elapses between the two events in this frame is

$$t'_{12} = \frac{1}{c}\sqrt{(c^2t_{12}^2 - l_{12}^2)} = \frac{s_{12}}{c}. \tag{34.6}$$

If two events occur in one and the same body, then the interval between them is always time-like, for the distance which the body moves between the two events cannot be greater than ct_{12}, since the velocity of the body cannot exceed c. So we have always

$$l_{12} < ct_{12}.$$

Let us now ask whether or not we can find a frame of reference in which two events occur at one and the same time. As before, we have for the frames K and K' $c^2t_{12}^2 - l_{12}^2 = c^2t_{12}'^2 - l_{12}'^2$. We want to have $t'_{12} = 0$, so that

$$s_{12}^2 = -l_{12}'^2 < 0.$$

Consequently the required frame can be found only for the case when the interval s_{12} between the two events is an imaginary number. Imaginary intervals are said to be *space-like*.

Thus if the interval between two events is space-like, there exists a frame of reference in which the two events occur simultaneously. The distance between the points where the events occur in this frame is

$$l'_{12} = \sqrt{(l_{12}^2 - c^2t_{12}^2)} = |s_{12}|. \tag{34.7}$$

The division of intervals into space- and time-like intervals is, because of their invariance, an absolute concept. This means that the time-like or space-like character of an interval is independent of the frame of reference.

Let us take some event O as our origin of time and space coordinates. In other words, in the four-dimensional system of coordinates, the axes of which are marked x, y, z, t, the world point of the event O is the origin of coordinates. Let us now consider what relation other

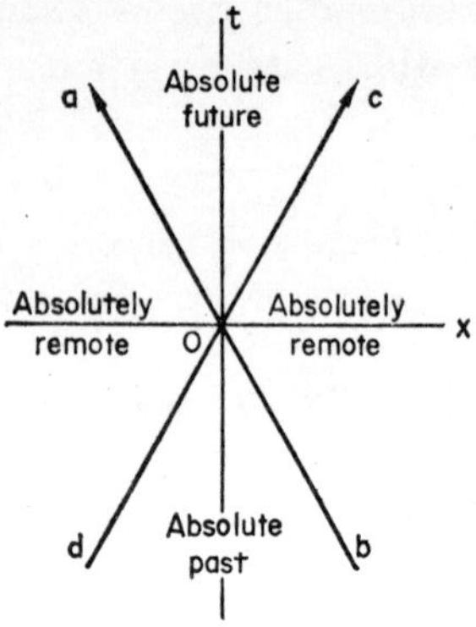

FIG. 28

events bear to the given event O. For visualisation, we shall consider only one space dimension and the time, marking them on two axes (Fig. 28). Uniform rectilinear motion of a particle, passing through $x = 0$ at $t = 0$, is represented by a straight line going through O and inclined to the t axis at an angle whose tangent is the velocity of the particle. Since the maximum possible velocity is c, there is a maximum angle which this line can make with the t axis. In Fig. 28 are shown the two lines representing the propagation of two signals (with the velocity of light) in opposite directions passing through the event O (i.e. going through $x = 0$ at $t = 0$). All lines representing the motion of particles can lie only in the regions aOc and dOb. On the lines ab and cd, $x = \pm ct$. First consider events whose world points lie within the region aOc. It is easy to show that for all the points of this region

$c^2t^2 - x^2 > 0$. In other words, the interval between any event in this region and the event O is time-like. In this region $t > 0$, i.e. all the events in this region occur "after" the event O. But two events which are separated by a time-like interval cannot occur simultaneously in any frame of reference. Consequently it is impossible to find a frame of reference in which any of the events in the region aOc occurred "before" the event O, i.e. at time $t < 0$. Thus all the events in the region aOc are future events relative to O in *all* frames of reference. Therefore this region can be called the *absolute future* relative to O.

In exactly the same way, all events in the region bOd are in the *absolute past* relative to O; i.e. events in this region occur before the event O in all frames of reference.

Finally, consider the regions dOa and cOb. The interval between any event in this region and the event O is space-like. These events occur at different points in space in every reference frame. Therefore these regions can be said to be *absolutely remote* relative to O. However, the concepts "simultaneous", "earlier", and "later" are relative for these events. For any event in these regions there exist frames of reference in which it occurs after the event O, frames in which it occurs earlier than O, and finally one reference frame in which it occurs simultaneously with O.

Note that if we consider all three space coordinates instead of just one, then instead of the two intersecting lines of Fig. 28 we would have a "cone" $x^2+y^2+z^2-c^2t^2 = 0$ in the four-dimensional coordinate system x, y, z, t, the axis of the cone coinciding with the t axis. (This cone is called the *light cone.*) The regions of "absolute future" and "absolute past" are then represented by the two interior portions of this cone.

Two events can be related causally to each other only if the interval between them is time-like; this follows immediately from the fact that no interaction can propagate with a velocity greater than the velocity of light. As we have just seen, it is precisely for these events that the concepts "earlier" and "later" have an absolute significance, which is a necessary condition for the concepts of cause and effect to have meaning.

§35. Proper time

Suppose that in a certain inertial reference frame K we observe a clock which is moving relative to us in an arbitrary manner. We also introduce an inertial frame K' moving relative to K with the velocity v of the clock at that time.

In the course of an infinitesimal time interval dt (as read by a clock in our rest frame) the moving clock goes a distance $\sqrt{(dx^2+dy^2+dz^2)}$. Let us ask what time interval dt' is indicated for this period by the moving clock. In the frame K' linked to the moving clock, the latter is at rest, i.e., $dx' = dy' = dz' = 0$. Because of the invariance of intervals

$$ds^2 = c^2\,dt^2 - dx^2 - dy^2 - dz^2 = c^2\,dt'^2,$$

from which

$$dt' = dt\sqrt{\left(1 - \frac{dx^2+dy^2+dz^2}{c^2\,dt^2}\right)}.$$

But

$$\frac{dx^2+dy^2+dz^2}{dt^2} = v^2,$$

where v is the velocity of the moving clock; therefore

$$dt' = \frac{ds}{c} = dt\sqrt{(1-v^2/c^2)}. \tag{35.1}$$

Integrating this expression, we can obtain the time interval indicated by the moving clock when the elapsed time according to a clock at rest is $t_2 - t_1$:

$$t'_2 - t'_1 = \int_{t_1}^{t_2} dt\sqrt{(1-v^2/c^2)}. \tag{35.2}$$

The time read by a clock moving with a given object is called the *proper time* for this object. Formulae (35.1) and (35.2) express the proper time in terms of the time for a frame of reference from which the motion is observed.

As we see from (35.1) or (35.2), the proper time of a moving object is always less than the corresponding interval in the rest frame. In other words, the moving clock goes more slowly than the one at rest.

Suppose a clock is moving in uniform rectilinear motion relative to an inertial frame K. A reference frame K' linked to this clock is also inertial. Then from the point of view of an observer in the frame K the clock in the frame K' falls behind. And conversely, from the point of view of the frame K', a clock in K lags. To convince ourselves that there is no contradiction, let us note the following. In order to establish that the clock in the frame K' lags behind that in the frame K, we must proceed in the following fashion. Suppose that at a certain moment the clock in K' passes by the clock in K, and at that moment the readings of the two clocks coincide. To compare the rates of the two clocks in K and K' we must once more compare the reading of the same moving clock in K' with a clock in K. But now we compare this clock with a *different* clock in K—with that past which the clock in K' goes at another time. Then we find that the clock in K' lags behind this other clock in K with which it is now being compared. We see that to compare the rates of clocks in two reference frames we require several clocks in one frame and one in the other, and that therefore this process is not symmetric with respect to the two frames. The clock that appears to lag is always the one which is being compared with different clocks in the other frame.

If we have two clocks, one of which describes a closed path returning to the starting point (the position of the clock which remained at rest), then the moving clock appears to lag relative to the one at rest. The converse reasoning, in which the moving clock would be considered to be at rest (and *vice versa*) is now impossible, since the clock describing a closed trajectory does not carry out a uniform rectilinear motion, so that a frame of reference linked to it will not be inertial. Since the laws of nature are the same only for inertial reference frames, the frames linked to the clock at rest (inertial frame) and to the moving clock (non-inertial) have different properties, and the argument which leads to the result that the clock at rest must lag is not valid.

§36. The Lorentz transformation

Our purpose is now to obtain the formulae of transformation from one inertial frame of reference to another, that is, formulae by means of which, knowing the coordinates x, y, z, t, of an event in a certain frame K, we can find the coordinates x', y', z', t' of the same event in another inertial frame K'.

In classical mechanics this question is resolved by the simple formulae for a Galileo transformation (3.1), (3.2). If the frame K' moves relative to the frame K along the common direction of the axes X and X', these formulae have the form

$$x = x'+Vt', \qquad y = y', \qquad z = z', \qquad t = t'. \tag{36.1}$$

This transformation, as was to be expected, does not satisfy the requirements of the theory of relativity; it does not leave the interval between events invariant.

We shall obtain the relativistic transformations precisely as a consequence of the requirement that they leave the interval between events invariant.

As we saw in §34, the interval between events can be looked on as the distance between the corresponding pair of world points in a four-dimensional system of coordinates. Consequently we may say that the required transformation must leave unchanged all distances in the four-dimensional x, y, z, ct space. But such transformations consist only of parallel displacements and rotations of the coordinate system. Of these the displacement of the coordinate system parallel to itself is of no interest, since it leads only to a shift in the origin of the space coordinates and a change in the time reference point. Thus the required transformation must be expressible mathematically as a rotation of the four-dimensional x, y, z, ct coordinate system.

Every rotation in the four-dimensional space can be resolved into six rotations, in the planes xy, zy, xz, tx, ty, tz (just as every rotation in ordinary space can be resolved into three rotations in the planes xy, zy, and xz). The first three of these rotations transform only the space coordinates; they correspond to the usual space rotations.

Let us consider a rotation in the tx plane; under this, the y and z coordinates do not change. This transformation must leave unchanged, in particular, the difference $(ct)^2-x^2$, the square of the "distance" from the point (ct, x) to the origin. The relation between the old and new coordinates in this transformation is given in general form by the formulae

$$x = x' \cosh \psi + ct' \sinh \psi, \quad ct = x' \sinh \psi + ct' \cosh \psi, \tag{36.2}$$

where ψ is the "angle of rotation"; a simple check shows that indeed $c^2t^2-x^2 = c^2t'^2-x'^2$. Formulae (36.2) differ from the usual formulae for transformation under a rotation of the axes in that trigonometric functions are replaced by hyperbolic functions. This is a manifestation of the difference between pseudo-Euclidean and Euclidean geometry.

We try to find the formula of transformation from an inertial reference frame K to a frame K' moving relative to K with velocity V along the x axis. In this case clearly only the coordinate x and the time t are subject to change. Therefore this transformation must have the form (36.2). Now it remains only to determine the angle ψ, which can depend only on the relative velocity V.†

Let us consider the motion, in the frame K, of the origin of the frame K'. Then $x' = 0$ and formulae (36.2) take the form

$$x = ct' \sinh \psi, \quad ct = ct' \cosh \psi,$$

or, dividing one by the other,

$$x/ct = \tanh \psi.$$

But x/t is clearly the velocity V of the frame K' relative to K. So

$$\tanh \psi = V/c.$$

From this

$$\sinh \psi = \frac{V/c}{\sqrt{[1-(V^2/c^2)]}}, \quad \cosh \psi = \frac{1}{\sqrt{[1-(V^2/c^2)]}}.$$

† Note to avoid confusion that we shall always use V to signify the constant relative velocity of two inertial frames, and v for the velocity of a moving particle, not necessarily constant.

Substituting in (36.2), we find

$$x = \frac{x'+Vt'}{\sqrt{[1-(V^2/c^2)]}}, \quad y = y', \quad z = z', \quad t = \frac{t'+(Vx'/c^2)}{\sqrt{[1-(V^2/c^2)]}}. \tag{36.3}$$

This is the required transformation formula. It is called the *Lorentz transformation*, and is of fundamental importance for what follows.

The inverse formulae, expressing x', y', z', t' in terms of x, y, z, t, are most easily obtained by changing V to $-V$ (since the frame K moves with velocity $-V$ relative to the frame K'). The same formulae can be obtained directly by solving equations (36.3) for x', y', z', t'.

It is easy to see from (36.3) that on making the transition to the limit $c \to \infty$ and classical mechanics, the formula for the Lorentz transformation actually goes over into the Galileo transformation.

For $V > c$ in formula (36.3) the coordinates x, t are imaginary; this corresponds to the fact that motion with a velocity greater than the velocity of light is impossible. Moreover, one cannot use a reference frame moving with the velocity of light—in that case the denominators in (36.3) would go to zero.

Suppose there is a rod at rest in the frame K, parallel to the x axis. Let its length, measured in this frame, be $\Delta x = x_2 - x_1$ (x_2 and x_1 are the coordinates of the two ends of the rod in the frame K). We now determine the length of this rod as measured in the frame K'. To do this we must find the coordinates of the two ends of the rod (x_2' and x_1') in this frame at one and the same time t'. From (36.3) we find

$$x_1 = \frac{x_1'+Vt'}{\sqrt{[1-(V^2/c^2)]}}, \quad x_2 = \frac{x_2'+Vt'}{\sqrt{[1-(V^2/c^2)]}}.$$

The length of the rod in the frame K' is $\Delta x' = x_2' - x_1'$; subtracting x_1 from x_2, we find

$$\Delta x = \frac{\Delta x'}{\sqrt{[1-(V^2/c^2)]}}.$$

The *proper length* of a rod is its length in a reference frame in which it is at rest. Let us denote it by $l_0 = \Delta x$, and the length of the rod in any other reference frame K' by l. Then

$$l = l_0\sqrt{[1-(V^2/c^2)]}. \tag{36.4}$$

Thus a rod has its greatest length in the reference frame in which it is at rest. Its length in a frame in which it moves with velocity V is decreased by the factor $\sqrt{[1-(V^2/c^2)]}$. This result of the theory of relativity is called the *Lorentz contraction.*

Since the transverse dimensions of a body do not change because of its motion, the volume $\mathcal{V}$ of a body decreases according to the similar formula

$$\mathcal{V} = \mathcal{V}_0\sqrt{[1-(V^2/c^2)]}, \tag{36.5}$$

where $\mathcal{V}_0$ is the *proper volume* of the body.

From the Lorentz transformation we can obtain anew the results already known to us concerning the proper time (§35). Suppose a clock to be at rest in the frame K'. We take two events occurring at one and the same point x', y', z' in space in the frame K'. The time between these events in the frame K' is $\Delta t' = t_2'-t_1'$. Now we find the time Δt which elapses between these two events in the frame K. From (36.3), we have

$$t_1 = \frac{t_1'+(Vx'/c^2)}{\sqrt{[1-(V^2/c^2)]}}, \quad t_2 = \frac{t_2'+(Vx'/c^2)}{\sqrt{[1-(V^2/c^2)]}},$$

or, subtracting one from the other,

$$t_2-t_1 = \Delta t = \frac{\Delta t'}{\sqrt{[1-(V^2/c^2)]}}$$

in complete agreement with (35.1).

§37. Transformation of velocities

In the preceding section we obtained formulae which enable us to find, from the coordinates of an event in one reference frame, the coordinates of the same event in a second reference frame. Now we find formulae relating the velocity of a material particle in one reference frame to its velocity in a second reference frame.

Let us suppose once again that the frame K' moves relative to the frame K with velocity V along the x axis. Let $v_x = \mathrm{d}x/\mathrm{d}t$ be the component of the particle velocity in the frame K and $v_x' = \mathrm{d}x'/\mathrm{d}t'$ the velocity component of the same particle in the frame K'. From (36.3), we have

$$\mathrm{d}x = \frac{\mathrm{d}x' + V\,\mathrm{d}t'}{\sqrt{[1-(V^2/c^2)]}}, \quad \mathrm{d}y = \mathrm{d}y', \quad \mathrm{d}z = \mathrm{d}z', \quad \mathrm{d}t = \frac{\mathrm{d}t' + (V\,\mathrm{d}x'/c^2)}{\sqrt{[1-(V^2/c^2)]}}.$$

Dividing the first three equations by the fourth and introducing the velocities

$$\mathbf{v} = \frac{\mathrm{d}\mathbf{r}}{\mathrm{d}t}, \quad \mathbf{v}' = \frac{\mathrm{d}\mathbf{r}'}{\mathrm{d}t'},$$

we find

$$v_x = \frac{v_x' + V}{1 + (v_x' V/c^2)}, \quad v_y = \frac{v_y' \sqrt{[1-(V^2/c^2)]}}{1 + (v_x' V/c^2)}, \quad v_z = \frac{v_z' \sqrt{[1-(V^2/c^2)]}}{1 + (v_x' V/c^2)}. \tag{37.1}$$

These formulae determine the transformation of velocities. They describe the law of composition of velocities in the theory of relativity. In the limiting case of $c \to \infty$, they go over into the formulae $v_x = v_x' + V$, $v_y = v_y'$, $v_z = v_z'$ of classical mechanics.

In the special case of motion of a particle parallel to the x axis, $v_x = v$, $v_y = v_z = 0$. Then $v_y' = v_z' = 0$, $v_x' = v'$, so that

$$v = \frac{v' + V}{1 + (v'V/c^2)}. \tag{37.2}$$

It is easy to convince oneself that the sum of two velocities each smaller than or equal to the velocity of light is again not greater than the light velocity.

Let us choose our coordinate axes so that the velocity of the particle at the given moment lies in the xy plane. Then the velocity of the particle in the frame K has components $v_x = v \cos\theta$, $v_y = v \sin\theta$, and in the frame K' $v_x' = v' \cos\theta'$, $v_y' = v' \sin\theta'$ (v, v', θ, θ' are the absolute values and the angles subtended with the x and x' axes, respectively in the frames K, K'). With the help of formulae (37.1), we then find

$$\tan\theta = \frac{v'\sin\theta'\sqrt{[1-(V^2/c^2)]}}{v'\cos\theta'+V}. \tag{37.3}$$

This formula describes the change in the direction of the velocity on transforming from one reference frame to another.

Let us consider a very important special case of this formula, namely, the deviation of light in transforming to a new reference frame—a phenomenon known as the *aberration of light*. In this case $v = v' = c$, so that the preceding formula goes over into

$$\tan\theta = \frac{\sqrt{[1-(V^2/c^2)]}}{V/c+\cos\theta'}\sin\theta'. \tag{37.4}$$

In the case $V \ll c$, we find from this formula, correct to terms of order V/c,

$$\tan\theta = \tan\theta'(1-V/c\cos\theta').$$

Introducing the angle $\Delta\theta = \theta'-\theta$ (the aberration angle), we find to the same order of accuracy

$$\Delta\theta = \frac{V}{c}\sin\theta', \tag{37.5}$$

which is the well-known elementary formula for the aberration of light.

§38. Four-vectors

The coordinates of an event (ct, x, y, z) can be considered as the components of a four-dimensional position vector (or, for short, a position four-vector) in a four-dimensional space. We shall denote its components by x^μ, where the index μ takes the values 0, 1, 2, 3, and

$$x^0 = ct, \quad x^1 = x, \quad x^2 = y, \quad x^3 = z.$$

The square of the "length" of the position four-vector is given by

$$(x^0)^2-(x^1)^2-(x^2)^2-(x^3)^2.$$

It does not change under any rotations of the four-dimensional coordinate system, in particular under Lorentz transformations.

In general a set of four quantities A^0, A^1, A^2, A^3 which transform like the components of the position four-vector† x^μ under transformations of the four-dimensional coordinate system is called a *four-dimensional vector* (*four-vector*) A^μ. Under Lorentz transformations,

$$A^0 = \frac{A'^0+(VA'^1/c)}{\sqrt{[1-(V^2/c^2)]}}, \quad A^1 = \frac{A'^1+(VA'^0/c)}{\sqrt{[1-(V^2/c^2)]}}, \quad A^2 = A'^2, \quad A^3 = A'^3. \tag{38.1}$$

The squared magnitude of any four-vector is defined analogously to the square of the radius four-vector:

$$(A^0)^2-(A^1)^2-(A^2)^2-(A^3)^2.$$

For convenience of notation, we introduce two "types" of components of four-vectors, denoting them by the symbols A^μ and A_μ, with superscripts and subscripts. These are related by

$$A_0 = A^0, \quad A_1 = -A^1, \quad A_2 = -A^2, \quad A_3 = -A^3. \tag{38.2}$$

The quantities A^μ are called the *contravariant*, and the A_μ the *covariant* components of the four-vector. The square of the four-vector then appears in the form

$$\sum_{\mu=0}^{3} A^\mu A_\mu = A^0A_0+A^1A_1+A^2A_2+A^3A_3.$$

Such sums are customarily written simply as $A^\mu A_\mu$, omitting the summation sign. The convention is that a summation is effected over any index which appears twice, and the summation sign is omitted. Of the pair of identical indices, one must be a superscript and the other a subscript. This convention for summation over "dummy" indices is very convenient and considerably simplifies the writing of formulae.

---

† We use Greek letters λ, μ, ν, ... for four-dimensional indices, taking values 0, 1, 2, 3.

In analogy to the square of a four-vector, one forms the *scalar product* of two different four-vectors:

$$A^{\mu}B_{\mu} = A^{0}B_{0}+A^{1}B_{1}+A^{2}B_{2}+A^{3}B_{3}.$$

It is clear that this can be written either as $A^{\mu}B_{\mu}$ or $A_{\mu}B^{\mu}$—the result is the same. In general one can switch upper and lower indices in any pair of dummy indices.†

The product $A^{\mu}B_{\mu}$ is a *four-scalar*—it is invariant under rotations of the four-dimensional coordinate system. This is easily verified directly, but it is also apparent beforehand (from the analogy with the square $A^{\mu}A_{\mu}$) from the fact that all four-vectors transform according to the same rule.

The component A^0 is called the *time component*, and A^1, A^2, A^3 the *space components* of the four-vector (in analogy to the position four-vector). The square of a four-vector can be positive, negative, or zero; such vectors are called *time-like*, *space-like*, and *null vectors*, respectively (again in analogy to the terminology for intervals).

Under purely spatial rotations (i.e. transformations not affecting the time axis) the three space components of the four-vector A^{μ} form a three-dimensional vector $\mathbf{A}$. The time component of the four-vector is a three-dimensional scalar (with respect to these transformations). In enumerating the components of a four-vector, we shall often write them as

$$A^{\mu} = (A^0, \mathbf{A}).$$

The covariant components of the same four-vector are $A_{\mu} = (A^0, -\mathbf{A})$, and the square of the four-vector is $A^{\mu}A_{\mu} = (A^0)^2-\mathbf{A}^2$. Thus, for the position four-vector,

$$x^{\mu} = (ct,\ \mathbf{r}), \quad x_{\mu} = (ct,\ -\mathbf{r}), \quad x^{\mu}x_{\mu} = c^2t^2-\mathbf{r}^2.$$

For three-dimensional vectors (with coordinates x, y, z) there is no need to distinguish between contra- and covariant components.

† In the literature the indices are often omitted on four-vectors, and their squares and scalar products are written as A^2, AB. We shall not use this notation in the present book.

Whenever this can be done without causing confusion, we shall write their components as A_i ($i = x, y, z$) using Latin letters for subscripts. In particular we shall assume a summation over x, y, z for any repeated index (for example, $\mathbf{A}\cdot\mathbf{B} = A_iB_i$).

A *four-dimensional tensor* (*four-tensor*) of rank two is a set of sixteen quantities $A^{\mu\nu}$, which under coordinate transformations transform like the products of components of two four-vectors. We similarly define four-tensors of higher rank.

The components of a four-tensor of rank two can be written in three forms: covariant, $A_{\mu\nu}$, contravariant, $A^{\mu\nu}$, and mixed, $A^{\mu}{}_{\nu}$ (where, in the last case, one should in general distinguish between $A^{\mu}{}_{\nu}$ and $A_{\mu}{}^{\nu}$, i.e. one should be careful about whether the first or the second is the subscript). The connection between the different types of components is determined from the general rule: raising or lowering a space index (1, 2, 3) changes the sign of the component, while raising or lowering the time index (0) does not. Thus:

$$A_{00} = A^{00}, \quad A_{01} = -A^{01}, \quad A_{11} = A^{11}, \ldots,$$

$$A_0{}^0 = A^{00}, \quad A_0{}^1 = A^{01}, \quad A^0{}_1 = -A^{01}, \quad A_1{}^1 = -A^{11}, \ldots.$$

Under purely spatial transformations, the nine quantities A^{11}, A^{12}, ... form a three-dimensional tensor. The three components A^{01}, A^{02}, A^{03} and the three components A^{10}, A^{20}, A^{30} constitute three-dimensional vectors, while the component A^{00} is a three-dimensional scalar.

A tensor $A^{\mu\nu}$ is said to be *symmetric* if $A^{\mu\nu} = A^{\nu\mu}$, and *antisymmetric* if $A^{\mu\nu} = -A^{\nu\mu}$. In an antisymmetric tensor, all the diagonal components (i.e. the components A^{00}, A^{11}, ...) are zero, since, for example, we must have $A^{00} = -A^{00}$. For a symmetric tensor $A^{\mu\nu}$, the mixed components $A^{\mu}{}_{\nu}$ and $A_{\nu}{}^{\mu}$ obviously coincide; in such cases we shall simply write A^{μ}_{ν}, putting the indices one above the other.

In every tensor equation, the two sides must contain identical and identically placed (i.e. above or below) free indices (as distinguished from dummy indices). The free indices in tensor equations can be shifted up or down, but this must be done simultaneously in all terms in the equation. Equating covariant and contravariant components of different tensors is "illegal"; such an equation, even if it happened by chance to be valid in a particular frame of reference, would be violated on going to another frame.

From the tensor components $A^{\mu\nu}$ one can form a scalar by taking the sum

$$A^{\mu}{}_{\mu} = A^{0}{}_{0}+A^{1}{}_{1}+A^{2}{}_{2}+A^{3}{}_{3}$$

(where, of course, $A^{\mu}{}_{\mu} = A_{\mu}{}^{\mu}$). This sum is called the *trace* of the tensor, and the operation for obtaining it is called *contraction.*

The formation of the scalar product of two four-vectors, considered earlier, is a contraction operation: it is the formation of the scalar $A^{\mu}B_{\mu}$ from the tensor $A^{\mu}B_{\nu}$. In general, contracting on any pair of indices reduces the rank of the tensor by 2. For example, $A^{\mu}{}_{\nu\lambda\mu}$ is a tensor of rank two, $A^{\mu}{}_{\nu}B^{\nu}$ is a four-vector, $A^{\mu\nu}{}_{\mu\nu}$ is a scalar, etc.

The unit four-tensor δ^{μ}_{ν} satisfies the condition that, for any four-vector A^{μ},

$$\delta^{\nu}_{\mu}A^{\mu} = A^{\nu}. \tag{38.3}$$

It is clear that the components of this tensor are

$$\delta^{\nu}_{\mu} = \begin{cases} 1, & \text{if} \quad \mu = \nu, \\ 0, & \text{if} \quad \mu \neq \nu. \end{cases} \tag{38.4}$$

Its trace is $\delta^{\mu}_{\mu} = 4$.

By raising the one index or lowering the other in δ^{ν}_{μ}, we can obtain the contra- or covariant tensor $g^{\mu\nu}$ or $g_{\mu\nu}$, which is called the *metric tensor*. The tensors $g^{\mu\nu}$ and $g_{\mu\nu}$ have identical components, which can be written as a matrix:

$$(g^{\mu\nu}) = (g_{\mu\nu}) = \begin{pmatrix} 1 & 0 & 0 & 0 \\ 0 & -1 & 0 & 0 \\ 0 & 0 & -1 & 0 \\ 0 & 0 & 0 & -1 \end{pmatrix} \tag{38.5}$$

(the index μ labels the rows, and ν the columns, in the order 0, 1, 2, 3). It is clear that

$$g_{\mu\nu}A^{\nu} = A_{\mu}, \quad g^{\mu\nu}A_{\nu} = A^{\mu}, \tag{38.6}$$

and this is the tensor form of the operation of lowering or raising indices.

Finally we consider some differential and integral operations of four-dimensional tensor analysis.

The four-gradient of a scalar ϕ is the four-vector

$$\frac{\partial \phi}{\partial x^\mu} = \left(\frac{1}{c}\frac{\partial \phi}{\partial t}, \nabla\phi\right).$$

Note that the derivatives should be regarded as the covariant components of the four-vector. For the differential of a scalar

$$\mathrm{d}\phi = \frac{\partial \phi}{\partial x^\mu}\,\mathrm{d}x^\mu$$

is also a scalar; from its form (scalar product of two four-vectors) our statement is clear.

In general, operators of differentiation with respect to the coordinates x^μ, $\partial/\partial x^\mu$, should be regarded as the covariant components of an operator four-vector. Thus the divergence of a four-vector, the expression $\partial A^\mu/\partial x^\mu$, in which the contravariant components of the four-vector A^μ are differentiated, is a scalar.

In three-dimensional space we can integrate over a volume, a surface, or a curve. There are four corresponding types of integrations in four-dimensional space: (1) over a curve in four-space, (2) over a (two-dimensional) surface, (3) over a hypersurface, i.e. a three-dimensional manifold, (4) over a four-dimensional volume.

In analogy to the theorems of Gauss and Stokes in three-dimensional vector analysis, there are theorems that enable one to transform four-dimensional integrals into one another. Of these we will need only the theorem for transforming an integral over a four-volume into an integral over a hypersurface.

The element of integration over a four-volume

$$\mathrm{d}\Omega = \mathrm{d}x^0\,\mathrm{d}x^1\,\mathrm{d}x^2\,\mathrm{d}x^3 = c\,\mathrm{d}t\,\mathrm{d}V \tag{38.7}$$

is a scalar; this is obvious from a comparison of the laws of transformation of time intervals (35.1) and of spatial volumes (36.5). The element of integration over a hypersurface $\mathrm{d}S^\mu$ is a four-vector equal in magnitude to the "area" of the hypersurface element and directed along the normal to the element (thus, the component $dS^0 = dx\,dy\,dz$, i.e. it is the three-dimensional volume element dV, the projection of the hypersurface element on the hyperplane $x^0 =$ constant).

The integral over a closed hypersurface can be transformed into an integral over the four-volume it encloses, by replacing the element of integration dS_μ by an operator:

$$dS_\mu \to d\Omega \frac{\partial}{\partial x^\mu}. \tag{38.8}$$

For example, for the integral of a vector A^μ we have

$$\oint A^\mu\, dS_\mu = \int \frac{\partial A^\mu}{\partial x^\mu}\, d\Omega. \tag{38.9}$$

This formula is the generalisation of Gauss' theorem.

CHAPTER 9

RELATIVISTIC MECHANICS

§39. Energy and momentum

Just as in classical mechanics, we shall start from the principle of least action in order to derive the relativistic equations of motion of particles. Let us start by finding the action integral for a free particle.

This integral must not depend on our particular choice of inertial reference frame, i.e. it must be invariant under Lorentz transformations. It follows that it must depend on a scalar. The integrand must be a first-order differential. But the only scalar of this kind that one can construct for a free particle is the interval ds, or ds multiplied by a constant characterising the particle. We shall denote this constant by $-mc$; the reason for this will be made clear in what follows.

So for a free particle the action must have the form

$$S = -mc \int_a^b \mathrm{d}s, \tag{39.1}$$

where $\int_a^b$ is an integral along the world line of the particle between two events—the presence of the particle at its initial and final positions at times t_1 and t_2.

By using (35.1) we can rewrite this expression as an integral over the time:

$$S = -mc^2 \int_{t_1}^{t_2} \sqrt{[1-(v^2/c^2)]}\, \mathrm{d}t.$$

Comparing this with the general definition (2.1)

$$S = \int_{t_1}^{t_2} L \, dt,$$

we see that the relativistic Lagrangian for a free particle is

$$L = -mc^2 \sqrt{[1-(v^2/c^2)]}. \tag{39.2}$$

For low velocities, in the non-relativistic limit, we can expand L in powers of v/c. Omitting higher-order terms, we get

$$L = -mc^2 + \tfrac{1}{2}mv^2.$$

A constant term in the Lagrangian has no effect on the equations of motion, and can be dropped. We then get back the classical expression $L = \frac{1}{2}mv^2$. We also see the significance of the constant m introduced in (39.1), which coincides with the mass of the particle.

By the *momentum* of a particle we mean the vector $\mathbf{p} = \partial L/\partial \mathbf{v}$. Differentiating (39.2), we find

$$\mathbf{p} = \frac{m\mathbf{v}}{\sqrt{[1-(v^2/c^2)]}}. \tag{39.3}$$

For small velocities ($v \ll c$) this expression goes over into the classical

$$\mathbf{p} = m\mathbf{v}.$$

The time derivative of the momentum is the force acting on the particle. Suppose the velocity of the particle changes only in direction, that is, suppose the force is directed perpendicular to the velocity. Then

$$\frac{d\mathbf{p}}{dt} = \frac{m}{\sqrt{[1-(v^2/c^2)]}} \frac{d\mathbf{v}}{dt}. \tag{39.4}$$

If the velocity changes only in magnitude, that is, if the force is parallel to the velocity, then

$$\frac{d\mathbf{p}}{dt} = \frac{m}{[1-(v^2/c^2)]^{3/2}} \frac{d\mathbf{v}}{dt}. \tag{39.5}$$

We see that the ratio of force to acceleration is different in the two cases.

According to the general definition (6.1), the *energy* E of the particle

$$E = \mathbf{p}\cdot\mathbf{v} - L. \tag{39.6}$$

Substituting the expressions (39.2) and (39.3) for L and $\mathbf{p}$, we find

$$E = \frac{mc^2}{\sqrt{[1-(v^2/c^2)]}}. \tag{39.7}$$

This very important formula shows, in particular, that in relativistic mechanics the energy of a free particle does not go to zero for $v = 0$, but rather takes on a finite value

$$E = mc^2. \tag{39.8}$$

This quantity is called thc *rest energy* of the particle.

For small velocities ($v/c \ll 1$), we have, expanding (39.7) in series in powers of v/c,

$$E \approx mc^2 + \tfrac{1}{2}mv^2,$$

which, except for the rest energy, is the classical expression for the kinetic energy of a particle.

We emphasise that, although we speak of a "particle", we have nowhere made use of the fact that it is "elementary". Thus the formulae are equally applicable to any composite body consisting of many particles, where by m we mean the total mass of the body and by v the velocity of its motion as a whole. In particular, formula (39.8) is valid for any body which is at rest as a whole. We call attention to the fact that in relativistic mechanics the energy of a free body (i.e. the energy of any closed system) is a completely definite quantity which is always positive and is directly related to the mass of the body. In this connection we recall that in classical mechanics the energy of a body is defined only to within an arbitrary additive constant, and can be either positive or negative.

The energy of a body at rest contains, in addition to the rest energies of its constituent particles, the kinetic energy of the particles and the energy of their interactions with one another. In other words, mc^2

is not equal to $\Sigma m_a c^2$ (where m_a are the masses of the particles), and so m is not equal to Σm_a. Thus in relativistic mechanics the law of conservation of mass does not hold: the mass of a composite body is not equal to the sum of the masses of its parts. Instead only the law of conservation of energy, in which the rest energies of the particles are included, is valid.

Squaring (39.3) and (39.7) and comparing the results, we get the following relation between the energy and momentum of a particle:

$$E^2/c^2 = p^2 + m^2c^2. \tag{39.9}$$

The energy expressed in terms of the momentum is called the Hamiltonian function H:

$$H = c\sqrt{(p^2+m^2c^2)}. \tag{39.10}$$

For low velocities, $p \ll mc$, and we have approximately

$$H \approx mc^2 + (p^2/2m),$$

i.e., except for the rest energy we get the familiar classical expression for the Hamiltonian.

From (39.3) and (39.7) we get the following relation between the energy, momentum, and velocity of a free particle:

$$\mathbf{p} = E\mathbf{v}/c^2. \tag{39.11}$$

For $v = c$, the momentum and energy of the particle become infinite. This means that a particle with mass m different from zero cannot move with the velocity of light. Nevertheless, in relativistic mechanics, particles of zero mass moving with the velocity of light can exist.† From (39.11) we have for such particles:

$$p = E/c. \tag{39.12}$$

The same formula also holds approximately for particles with non-zero mass in the *ultra-relativistic* case, when the particle energy E is large compared with its rest energy mc^2.

---

† Examples are light quanta and neutrinos.

§40. Four-momentum

The arguments of the previous section still leave open the question of the transformation of the energy and momentum of a particle when we change from one reference frame to another. To answer this question we must make clear the four-dimensional nature of these quantities.

From the usual three-dimensional velocity vector $\mathbf{v}$ of a particle we can also form a four-vector. This *velocity four-vector* is

$$u^\mu = \mathrm{d}x^\mu/\mathrm{d}s. \tag{40.1}$$

Expressing the element of interval $\mathrm{d}s$ in terms of the time differential $\mathrm{d}t$ according to (35.1) we can write

$$u^\mu = \frac{1}{\sqrt{[1-(v^2/c^2)]}}\frac{\mathrm{d}x^\mu}{c\,\mathrm{d}t}.$$

From this we see that the components of this four-vector are

$$u^\mu = \left(\frac{1}{\sqrt{[1-(v^2/c^2)]}}, \frac{\mathbf{v}}{c\sqrt{[1-(v^2/c^2)]}}\right). \tag{40.2}$$

The components of the four-velocity are not independent. Noting that $\mathrm{d}x_\mu\,\mathrm{d}x^\mu = \mathrm{d}s^2$, we find

$$u_\mu u^\mu = 1. \tag{40.3}$$

From the geometrical point of view u^μ is a unit four-vector tangent to the world line of the particle.

The *four-momentum* of a particle is the four-vector

$$p^\mu = mcu^\mu. \tag{40.4}$$

Taking the components of the four-velocity from (40.2) and comparing with the expressions (39.3) and (39.7), we see that the components of the four-momentum are

$$p^\mu = (E/c, \mathbf{p}). \tag{40.5}$$

Thus in relativistic mechanics momentum and energy are components of a single four-vector. From this we immediately get the formulae for transformation of these quantities. Substituting the expressions (40.5) in the general law (38.1) for transformation of four-vectors, we find

$$p_x = \frac{p_x'+(VE'/c^2)}{\sqrt{[1-(V^2/c^2)]}}, \quad p_y = p_y', \quad p_z = p_z', \quad E = \frac{E'+Vp_x'}{\sqrt{[1-(V^2/c^2)]}}, \tag{40.6}$$

where p_x, p_y, p_z are the components of the three-momentum $\mathbf{p}$.

From the definition (40.4) of the four-momentum and the identity (40.3) we have for the square of the four-momentum of a free particle

$$p_\mu p^\mu = m^2c^2. \tag{40.7}$$

Substituting the components p^μ from (40.5), we arrive at the relation (39.9).

§41. Decay of particles

Let us consider the spontaneous decay of a body of mass M into two parts with masses m_1 and m_2. The law of conservation of energy in the decay, applied in the frame of reference in which the body is at rest, gives†

$$M = E_{10}+E_{20}, \tag{41.1}$$

where E_{10} and E_{20} are the energies of the emerging particles. Since $E_{10} > m_1$ and $E_{20} > m_2$, the equality (41.1) can be satisfied only if $M > m_1+m_2$, i.e. a body can disintegrate spontaneously into parts the sum of whose masses is less than the mass of the body. On the other

† In §§41, 42 we put $c = 1$. In other words the velocity of light is taken as the unit of velocity (so that the dimensions of length and time become the same). This choice is a natural one in relativistic mechanics and greatly simplifies the writing of formulae. However, in this book (which also contains a considerable amount of non-relativistic theory) we shall not usually use this system of units, and will explicitly indicate when we do.

If c has been put equal to unity in formulae, it is easy to convert back to ordinary units: the velocity of light is introduced to assure correct dimensions.

hand, if $M < m_1+m_2$, the body is stable (with respect to the particular decay) and does not decay spontaneously. To cause the decay in this case, we would have to supply to the body from outside an amount of energy at least equal to its *binding energy* (m_1+m_2-M).

Momentum as well as energy must be conserved in the decay process. Since the initial momentum of the body was zero, the sum of the momenta of the emerging particles must be zero: $\mathbf{p}_{10}+\mathbf{p}_{20} = 0$. Consequently $p_{10}^2 = p_{20}^2$, or

$$E_{10}^2 - m_1^2 = E_{20}^2 - m_2^2. \tag{41.2}$$

The two equations (41.1) and (41.2) uniquely determine the energies of the emerging particles:

$$E_{10} = \frac{M^2+m_1^2-m_2^2}{2M}, \quad E_{20} = \frac{M^2-m_1^2+m_2^2}{2M}. \tag{41.3}$$

In a certain sense the inverse of this problem is the calculation of the total energy M of two colliding particles in the frame of reference in which their total momentum is zero (the *centre-of-mass system* or *C system*). The computation of this quantity gives a criterion for the possible occurrence of various inelastic collision processes, accompanied by a change in state of the colliding particles, or the "creation" of new particles. A process of this type can occur only if the sum of the masses of the "reaction products" does not exceed M.

Suppose that in the initial reference frame (the *laboratory system* or *L system*) a particle with mass m_1 and energy E_1 collides with a particle of mass m_2 which is at rest. The total energy of the two particles is

$$E = E_1+E_2 = E_1+m_2,$$

and their total momentum is $\mathbf{p} = \mathbf{p}_1+\mathbf{p}_2 = \mathbf{p}_1$. Considering the two particles together as a single composite system, we find the velocity of its motion as a whole from (39.11):

$$\mathbf{V} = \frac{\mathbf{p}}{E} = \frac{\mathbf{p}_1}{E_1+m_2}. \tag{41.4}$$

This quantity is the velocity of the C system with respect to the L system.

However, in determining the mass M, there is no need to transform from one reference frame to the other. Instead we can make direct use of formula (39.9), which is applicable to the composite system just as it is to each particle individually. We thus have

$$M^2 = E^2 - p^2 = (E_1+m_2)^2 - (E_1^2 - m_1^2),$$

from which

$$M^2 = m_1^2 + m_2^2 + 2m_2 E_1. \tag{41.5}$$

PROBLEM

Determine the maximum energy which can be carried off by one of the decay particles, when a particle of mass M at rest decays into three particles with masses m_1, m_2 and m_3.

SOLUTION. The particle m_1 has its maximum energy if the system of the other two particles m_2 and m_3 has the least possible mass; the latter is equal to the sum m_2+m_3 (and corresponds to the case where the two particles move together with the same velocity). Having thus reduced the problem to the decay of a body into two parts we obtain from (41.3)

$$E_{1\,\max} = \frac{M^2 + m_1^2 - (m_2+m_3)^2}{2M}.$$

§42. Elastic collisions of particles

Let us consider, from the point of view of relativistic mechanics, the *elastic collision* of particles. We denote the momenta and energies of the two colliding particles (with masses m_1 and m_2) by $\mathbf{p}_1$, E_1 and $\mathbf{p}_2$, E_2; we use primes for the corresponding quantities after collision. The laws of conservation of momentum and energy in the collision can be written together as the equation for conservation of the four-momentum:

$$p_1^\mu + p_2^\mu = p_1'^\mu + p_2'^\mu. \tag{42.1}$$

From this four-vector equation we construct invariant relations which will be helpful in further computations. To do this we rewrite (42.1) in the form

$$p_1^\mu + p_2^\mu - p_1'^\mu = p_2'^\mu,$$

and square both sides (i.e. we write the scalar product of each side with itself). Noting that the squares of the four-momenta p_1^μ and $p_1'^\mu$ are equal to m_1^2, and the squares of p_2^μ and $p_2'^\mu$ are equal to m_2^2, we get

$$m_1^2+p_{1\mu}p_2^\mu-p_{1\mu}p_1'^\mu-p_{2\mu}p_1'^\mu = 0. \tag{42.2}$$

Similarly, squaring the equation $p_1^\mu+p_2^\mu-p_2'^\mu = p_1'^\mu$, we get

$$m_2^2+p_{1\mu}p_2^\mu-p_{2\mu}p_2'^\mu-p_{1\mu}p_2'^\mu = 0. \tag{42.3}$$

Let us consider the collision in a reference frame (the L system) in which one of the particles (m_2) was at rest before the collision. Then $\mathbf{p}_2 = 0$, $E_2 = m_2$, and the scalar products appearing in (42.2) are

$$\left.\begin{aligned} p_{1\mu}p_2^\mu &= E_1m_2, \\ p_{2\mu}p_1'^\mu &= m_2E_1', \\ p_{1\mu}p_1'^\mu &= E_1E_1'-\mathbf{p}_1\cdot\mathbf{p}_1' = E_1E_1'-p_1p_1'\cos\theta_1, \end{aligned}\right\} \tag{42.4}$$

where θ_1 is the angle of scattering of the incident particle m_1. Substituting these expressions in (42.2), we get

$$\cos\theta_1 = \frac{E_1'(E_1+m_2)-E_1m_2-m_1^2}{p_1p_1'}. \tag{42.5}$$

Similarly, we find from (42.3)

$$\cos\theta_2 = \frac{(E_1+m_2)(E_2'-m_2)}{p_1p_2'}, \tag{42.6}$$

where θ_2 is the angle between the transferred momentum $\mathbf{p}_2'$ and the momentum of the incident particle $\mathbf{p}_1$. These formulae relate the angles of scattering of the two particles in the L system to the changes in their energy in the collision.

We note that if $m_1 > m_2$, i.e. if the incident particle is heavier than the target particle, the scattering angle θ_1 cannot exceed a certain maximum value. It is easy to find by elementary computations that this value is given by the equation

$$\sin\theta_{1\,\max} = m_2/m_1, \tag{42.7}$$

which coincides with the familiar classical result (14.8).

Formulae (42.5), (42.6) simplify in the case when the incident particle has zero mass: $m_1 = 0$, and correspondingly $p_1 = E_1$, $p_1' = E_1'$. For this case let us write the formula for the energy of the incident particle after the collision, expressed in terms of its angle of deflection:

$$E_1' = \frac{m_2}{1-\cos\theta_1+m_2/E_1}. \tag{42.8}$$

Let us now turn once again to the general case of collision of particles of arbitrary mass. The collision is most simply treated in the C system. Designating quantities in this system by the additional subscript 0, we have $\mathbf{p}_{10} = -\mathbf{p}_{20} \equiv \mathbf{p}_0$. From the conservation of momentum, during the collision the momenta of the two particles merely rotate, remaining equal in magnitude and opposite in direction. From the conservation of energy, the value of each of the momenta remains unchanged.

Let χ be the angle of scattering in the C system—the angle through which the momenta $\mathbf{p}_{10}$ and $\mathbf{p}_{20}$ are rotated by the collision. This quantity completely determines the scattering process in the C system, and therefore also in any other frame of reference. It is also convenient in describing the collision in the L system and serves as the single parameter which remains undetermined after the laws of conservation of momentum and energy are applied.

We express the final energies of the two particles in the L system in terms of this parameter. To do this we return to (42.2), but this time write out the product $p_{1\mu}p_1'^{\mu}$ in the C system:

$$p_{1\mu}p_1'^{\mu} = E_{10}E_{10}'-\mathbf{p}_{10}\cdot\mathbf{p}_{10}' = E_{10}^2-p_0^2\cos\chi = p_0^2(1-\cos\chi)+m_1^2$$

(in the C system the energies of the particles do not change in the collision: $E_{10}' = E_{10}$). We write out the other two products in the L system, i.e. we use (42.4). As a result we get $E_1'-E_1 = -(p_0^2/m_2)\times$ $\times(1-\cos\chi)$. We must still express p_0^2 in terms of quantities referring to the L system. This is easily done by equating the values of the invariant $p_{1\mu}p_2^{\mu}$ in the L and C systems:

$$E_{10}E_{20}-\mathbf{p}_{10}\cdot\mathbf{p}_{20} = E_1 m_2,$$

or

$$\sqrt{[(p_0^2+m_1^2)(p_0^2+m_2^2)]} = E_1 m_2 - p_0^2.$$

Solving the equation for p_0^2, we get

$$p_0^2 = \frac{m_2^2(E_1^2-m_1^2)}{m_1^2+m_2^2+2m_2E_1}. \qquad (42.9)$$

Also using the conservation law $E_1+m_2 = E_1'+E_2'$, we finally get

$$E_1-E_1' = E_2'-m_2 = \frac{m_2(E_1^2-m_1^2)}{m_1^2+m_2^2+2m_2E_1}(1-\cos\chi). \qquad (42.10)$$

This expression represents the energy lost by the first particle and transferred to the second particle. The maximum energy transfer occurs for $\chi = \pi$, and is equal to

$$E'_{2\,\max}-m_2 = E_1-E'_{1\,\min} = \frac{2m_2(E_1^2-m_1^2)}{m_1^2+m_2^2+2m_2E_1}. \qquad (42.11)$$

The ratio of the minimum kinetic energy of the incident particle after collision to its initial kinetic energy is:

$$\frac{E'_{1\,\min}-m_1}{E_1-m_1} = \frac{(m_1-m_2)^2}{m_1^2+m_2^2+2m_2E_1}. \qquad (42.12)$$

In the limiting case of low velocities (when $E \approx m+\frac{1}{2}mv^2$), this ratio tends to a constant limit, equal to

$$\left(\frac{m_1-m_2}{m_1+m_2}\right)^2.$$

In the opposite limit of large energies E_1, the ratio (42.12) tends to zero; the quantity $E'_{1\,\min}$ tends to a constant limit. This limit is

$$E'_{1\,\min} = \frac{m_1^2+m_2^2}{2m_2}.$$

Let us assume that $m_2 \gg m_1$, i.e. the mass of the incident particle is small compared with the mass of the particle at rest. According to classical mechanics the light particle could transfer only a negligible part of its energy (see §14). This is not the case in relativistic mechanics. From formula (42.12) we see that for sufficiently large energies E_1 the fraction of the energy transferred can reach the order of unity. For this it is not sufficient that the velocity of m_1 be of order 1, but one must have $E_1 \sim m_2$, i.e. the light particle must have an energy of the order of the rest energy of the heavy particle.

A similar situation occurs for $m_2 \ll m_1$, i.e. when a heavy particle is incident on a light one. Here too, according to classical mechanics, the energy transfer would be insignificant. The fraction of the energy transferred begins to be significant only for energies $E_1 \sim m_1^2/m_2$. We note that we are not talking simply of velocities of the order of the light velocity, but of energies large compared with m_1, i.e. we are dealing with the ultra-relativistic case.

PROBLEMS

PROBLEM 1. For two particles of equal mass, find the separation angle after collision, in the laboratory system.

SOLUTION. Taking the squares of both sides of (42.1), we get

$$p_{1\mu}p_2^{\mu} = p'_{1\mu}p_2'^{\mu}$$

and, after writing out the products of the four-momenta in the L system,

$$E_1 m_2 = E_1'E_2' - p_1'p_2' \cos\Theta,$$

where Θ is the separation angle (the angle between $\mathbf{p}_1'$ and $\mathbf{p}_2'$). For particles of equal mass ($m_1 = m_2 \equiv m$), we substitute

$$p_1' = \sqrt{(E_1'^2 - m^2)}, \qquad p_2' = \sqrt{(E_2'^2 - m^2)},$$

and using the conservation of energy ($E_1 + m = E_1' + E_2'$) we obtain

$$\cos\Theta = \sqrt{\frac{(E_1'-m)(E_2'-m)}{(E_1'+m)(E_2'+m)}}\,.$$

The angle Θ varies from $\frac{1}{2}\pi$ (for $E_1' \to m$ or $E_2' \to m$) to a minimum value $\Theta_{\min}$, which is reached when $E_1' = E_2'$:

$$\cos\Theta_{\min} = \frac{E_1 - m}{E_1 + 3m}\,.$$

PROBLEM 2. For the collision of two particles of equal mass m, express E_1', E_2' and χ in terms of the scattering angle θ_1 in the laboratory system.

SOLUTION. Substituting $p_1 = \sqrt{(E_1^2-m^2)}$, $p_1' = \sqrt{(E_1'^2-m^2)}$ in (42.5) and solving the equation for E_1', we find

$$E_1' = \frac{E_1+m+(E_1-m)\cos^2\theta_1}{E_1+m-(E_1-m)\cos^2\theta_1}$$

and so

$$E_2' = E_1+m-E_1' = m+\frac{(E_1^2-m^2)\sin^2\theta_1}{2m+(E_1-m)\sin^2\theta_1}.$$

Comparing with the expression for E_1' in terms of χ:

$$E_1' = E_1-\tfrac{1}{2}(E_1-m)(1-\cos\chi)$$

(from (42.10)), we find the angle of scattering in the centre-of-mass system:

$$\cos\chi = \frac{2m-(E_1+3m)\sin^2\theta_1}{2m+(E_1+m)\sin^2\theta_1}.$$

PART II

ELECTRODYNAMICS

CHAPTER 10

CHARGES IN ELECTROMAGNETIC FIELDS

§43. Four-potential of a field

The interaction of particles can be described with the help of the concept of a *field* of force. Namely, instead of saying that one particle acts on another, we may say that the particle creates a field around itself; a certain force then acts on every other particle located in this field. In classical mechanics, the field is merely a mode of description of the physical phenomenon—the interaction of particles. In the theory of relativity, because of the finite velocity of propagation of interactions, the situation is changed fundamentally. The forces acting on a particle at a given moment are not determined by the positions at that same moment. A change in the position of one of the particles influences other particles only after the lapse of a certain time interval. This means that the field itself acquires physical reality. We cannot speak of a direct interaction of particles located at a distance from one another. Interactions can occur at any one moment only between neighbouring points in space (contact interaction). Therefore we must speak of the interaction of the one particle with the field, and of the subsequent interaction of the field with the second particle.

The second part of this book is devoted to the theory of electromagnetic fields. We start by studying the interaction of a particle with a given field.

For a particle moving in a given electromagnetic field, the action is made up of two parts: the action (39.1) for the free particle, and a term describing the interaction of the particle with the field. The latter term must contain quantities characterising the particle and quantities characterising the field.

It turns out† that the properties of a particle with respect to interaction with the electromagnetic field are determined by a single parameter—the *charge* e of the particle, which can be either positive or negative (or equal to zero). The properties of the field are characterised by a four-vector A_μ, the *four-potential*, whose components are functions of the coordinates and time. These quantities appear in the action function in the term

$$-\frac{e}{c}\int_a^b A_\mu \, dx^\mu,$$

where the functions A_μ are taken at points on the world line of the particle. The factor $1/c$ has been introduced for convenience. It should be pointed out that, so long as we have no formulae relating the charge or the potentials with already known quantities, the units for measuring these new quantities can be chosen arbitrarily. (We shall return to this question in §53.)

Thus the action function for a charge in an electromagnetic field has the form

$$S = \int_a^b \left(-mc \, ds - \frac{e}{c} A_\mu \, dx^\mu\right). \tag{43.1}$$

† The assertions which follow should be regarded as being essentially the consequence of experimental data. The form of the action for a particle in an electromagnetic field cannot be fixed on the basis of general considerations alone, such as, for example, the requirement of relativistic invariance. The latter would permit, for example, the occurrence in formula (43.1) of a term of the form $\int A \, ds$, where A is a scalar function.

To avoid any misunderstanding, we repeat that we are considering classical (and not quantum) theory, and therefore do not include effects which are related to the spins of particles.

The three space components of the four-vector A^μ form a three-dimensional vector **A** called the *vector potential* of the field. The time component is called the *scalar potential*; we denote it by $A^0 = \phi$. Thus

$$A^\mu = (\phi, \mathbf{A}). \tag{43.2}$$

Therefore the action integral can be written in the form

$$S = \int_a^b \left(-mc\,\mathrm{d}s + \frac{e}{c}\mathbf{A}\cdot\mathrm{d}\mathbf{r} - e\phi\,\mathrm{d}t\right).$$

Introducing $\mathrm{d}\mathbf{r}/\mathrm{d}t = \mathbf{v}$, the particle velocity, and changing to an integration over t,

$$S = \int_{t_1}^{t_2} \left(-mc^2\sqrt{[1-(v^2/c^2)]} + \frac{e}{c}\mathbf{A}\cdot\mathbf{v} - e\phi\right)\mathrm{d}t. \tag{43.3}$$

The integrand is just the Lagrangian for a charge in an electromagnetic field:

$$L = -mc^2\sqrt{[1-(v^2/c^2)]} + \frac{e}{c}\mathbf{A}\cdot\mathbf{v} - e\phi. \tag{43.4}$$

This function differs from the Lagrangian for a free particle by the terms $(e/c)\,\mathbf{A}\cdot\mathbf{v} - e\phi$, which describe the interaction of the charge with the field.

The derivative $\partial L/\partial\mathbf{v}$ is the generalised momentum of the particle; we denote it by **P**:

$$\mathbf{P} = \frac{m\mathbf{v}}{\sqrt{[1-(v^2/c^2)]}} + \frac{e}{c}\mathbf{A} = \mathbf{p} + \frac{e}{c}\mathbf{A}. \tag{43.5}$$

Here we have denoted by **p** the ordinary momentum of the particle, which we shall refer to simply as its momentum.

From the Lagrangian we can find the Hamiltonian function for a particle in a field from the general formula†

$$\mathcal{H} = \mathbf{v}\cdot\frac{\partial L}{\partial \mathbf{v}} - L.$$

Substituting (43.4), we get

$$\mathcal{H} = \frac{mc^2}{\sqrt{[1-(v^2/c^2)]}} + e\phi. \tag{43.6}$$

However, the Hamiltonian must be expressed not in terms of the velocity, but rather in terms of the generalised momentum of the particle. From (43.5) and (43.6) it is clear that the relation between $\mathcal{H}-e\phi$ and $\mathbf{P}-(e/c)\mathbf{A}$ is the same as the relation between $\mathcal{H}$ and $\mathbf{p}$ in the absence of the field, i.e.

$$\left(\frac{\mathcal{H}-e\phi}{c}\right)^2 = m^2c^2 + \left(\mathbf{P}-\frac{e}{c}\mathbf{A}\right)^2. \tag{43.7}$$

For low velocities, i.e. for classical mechanics, the Lagrangian (43.4) goes over into

$$L = \tfrac{1}{2}mv^2 + \frac{e}{c}\mathbf{A}\cdot\mathbf{v} - e\phi. \tag{43.8}$$

In this approximation

$$\mathbf{p} = m\mathbf{v} = \mathbf{P} - \frac{e}{c}\mathbf{A},$$

and we find the following expression for the Hamiltonian:

$$\mathcal{H} = \frac{1}{2m}\left(\mathbf{P}-\frac{e}{c}\mathbf{A}\right)^2 + e\phi. \tag{43.9}$$

§44. Equations of motion of a charge in a field

A charge located in a field not only is subjected to a force exerted by the field, but also in turn acts on the field, changing it. However, if the charge e is not large, the action of the charge on the field can be

† In this part of the book we shall use script letters $\mathcal{E}$ and $\mathcal{H}$ (instead of E and H) for the energy and the Hamiltonian, in order to avoid confusion with the field strengths.

neglected.† In this case, when considering the motion of the charge in a given field, we may assume that the field itself does not depend on the coordinates or the velocity of the charge.

The equations of motion of a charge in a given electromagnetic field are given by the Lagrange equation

$$\frac{\mathrm{d}}{\mathrm{d}t}\left(\frac{\partial L}{\partial \mathbf{v}}\right) = \frac{\partial L}{\partial \mathbf{r}}, \tag{44.1}$$

where L is given by formula (43.4).

The derivative $\partial L/\partial \mathbf{v}$ is the generalised momentum of the particle (43.5). Further, we write

$$\frac{\partial L}{\partial \mathbf{r}} \equiv \nabla L = \frac{e}{c}\operatorname{grad} \mathbf{A}\cdot\mathbf{v} - e \operatorname{grad} \phi.$$

But from a formula of vector analysis

$$\operatorname{grad}(\mathbf{a}\cdot\mathbf{b}) = (\mathbf{a}\cdot\nabla)\mathbf{b} + (\mathbf{b}\cdot\nabla)\mathbf{a} + \mathbf{b}\times\operatorname{curl}\mathbf{a} + \mathbf{a}\times\operatorname{curl}\mathbf{b},$$

where $\mathbf{a}$ and $\mathbf{b}$ are two arbitrary vectors. Applying this formula to $\mathbf{A}\cdot\mathbf{v}$, and remembering that differentiation with respect to $\mathbf{r}$ is carried out for constant $\mathbf{v}$, we find

$$\frac{\partial L}{\partial \mathbf{r}} = \frac{e}{c}(\mathbf{v}\cdot\nabla)\mathbf{A} + \frac{e}{c}\mathbf{v}\times\operatorname{curl}\mathbf{A} - e\operatorname{grad}\phi.$$

So the Lagrange equation has the form

$$\frac{\mathrm{d}}{\mathrm{d}t}\left(\mathbf{p} + \frac{e}{c}\mathbf{A}\right) = \frac{e}{c}(\mathbf{v}\cdot\nabla)\mathbf{A} + \frac{e}{c}\mathbf{v}\times\operatorname{curl}\mathbf{A} - e\operatorname{grad}\phi.$$

But the total differential $(\mathrm{d}\mathbf{A}/\mathrm{d}t)\,\mathrm{d}t$ consists of two parts: the change $(\partial\mathbf{A}/\partial t)\,\mathrm{d}t$ of the vector potential with time at a fixed point in space, and the change due to motion from one point in space to another at

† The condition for smallness of the charge in this sense is that the radiation damping forces arising from its motion (cf. the discussion in §84) be small.

distance $\mathbf{dr}$. This second part is equal to $(\mathbf{dr}\cdot\nabla)\mathbf{A}$. Thus

$$\frac{\mathrm{d}\mathbf{A}}{\mathrm{d}t} = \frac{\partial\mathbf{A}}{\partial t} + (\mathbf{v}\cdot\nabla)\mathbf{A}.$$

Substituting this in the previous equation, we find

$$\frac{\mathrm{d}\mathbf{p}}{\mathrm{d}t} = -\frac{e}{c}\frac{\partial\mathbf{A}}{\partial t} - e\,\mathrm{grad}\,\phi + \frac{e}{c}\,\mathbf{v}\times\mathrm{curl}\,\mathbf{A}. \tag{44.2}$$

This is the equation of motion of a particle in an electromagnetic field. On the left-hand side stands the derivative of the particle's momentum with respect to the time. Therefore the expression on the right of (44.2) is the force exerted on the charge in an electromagnetic field. We see that this force consists of two parts. The first part (first and second terms on the right-hand side of (44.2)) does not depend on the velocity of the particle. The second part (third term) depends on the velocity, being proportional to the velocity and perpendicular to it.

The force of the first type, per unit charge, is called the *electric field intensity*; we denote it by $\mathbf{E}$. So, by definition,

$$\mathbf{E} = -\frac{1}{c}\frac{\partial\mathbf{A}}{\partial t} - \mathrm{grad}\,\phi. \tag{44.3}$$

The coefficient of $\mathbf{v}/c$ in the force of the second type, per unit charge, is called the *magnetic field intensity*. We designate it by $\mathbf{H}$. So, by definition,

$$\mathbf{H} = \mathrm{curl}\,\mathbf{A}. \tag{44.4}$$

The equation of motion of a charge in an electromagnetic field can now be written as

$$\frac{\mathrm{d}\mathbf{p}}{\mathrm{d}t} = e\mathbf{E} + \frac{e}{c}\,\mathbf{v}\times\mathbf{H}. \tag{44.5}$$

The expression on the right is called the *Lorentz force*. The first term (the force which the electric field exerts on the charge) does not depend on the velocity of the charge, and is along the direction of $\mathbf{E}$. The second part (the force exerted by the magnetic field on the charge) is proportional to the velocity of the charge and is directed perpendicular to the velocity and to the magnetic field **H**.

For velocities small compared with the velocity of light, the momentum **p** is approximately equal to its classical expression $m\mathbf{v}$, and the equation of motion (44.5) becomes

$$m\frac{d\mathbf{v}}{dt} = e\mathbf{E} + \frac{e}{c}\mathbf{v}\times\mathbf{H}. \tag{44.6}$$

Next we calculate the rate of change of the kinetic energy of the particle† with time, i.e. the derivative

$$\frac{d\mathscr{E}_{\text{kin}}}{dt} = \frac{d}{dt}\left(\frac{mc^2}{\sqrt{[1-(v^2/c^2)]}}\right).$$

It is easy to check that

$$\frac{d\mathscr{E}_{\text{kin}}}{dt} = \mathbf{v}\cdot\frac{d\mathbf{p}}{dt}.$$

Substituting $d\mathbf{p}/dt$ from (44.5) and noting that $\mathbf{v}\times\mathbf{H}\cdot\mathbf{v} = 0$, we have

$$\frac{d\mathscr{E}_{\text{kin}}}{dt} = e\mathbf{E}\cdot\mathbf{v}. \tag{44.7}$$

The expression on the right-hand side is the work done by the field on the particle per unit time. Work is done on the charge only by the electric field; the magnetic field does no work on a charge moving in it, since the force which the magnetic field exerts on a charge is always perpendicular to the velocity of the charge.

In §5 it was noted that the equations of classical mechanics are invariant with respect to a change in sign of the time. It is easy to see that this is also valid for the electromagnetic field in the theory of relativity. In this case, however, in addition to changing t into $-t$, we must reverse the sign of the magnetic field. It is easy to see that the

† By "kinetic" we mean the energy (39.7), which includes the rest energy.

equations of motion (44.5) are not altered if we make the changes

$$t \to -t, \quad \mathbf{E} \to \mathbf{E}, \quad \mathbf{H} \to -\mathbf{H}. \tag{44.8}$$

According to (44.3) and (44.4), this does not change the scalar potential, while the vector potential changes sign:

$$\phi \to \phi, \quad \mathbf{A} \to -\mathbf{A}, \tag{44.9}$$

Thus, if a certain motion is possible in an electromagnetic field, then the reversed motion is possible in a field in which the direction of **H** is reversed.

PROBLEM

Express the acceleration of a particle in terms of its velocity and the electric and magnetic field intensities.

SOLUTION. Substitute in the equation of motion (44.5) $\mathbf{p} = \mathbf{v}\mathscr{E}_{\text{kin}}/c^2$, and take the expression for $d\mathscr{E}_{\text{kin}}/dt$ from (44.7). The result is

$$\dot{\mathbf{v}} = \frac{e}{m}\sqrt{[1-(v^2/c^2)]}\left\{\mathbf{E}+\frac{1}{c}\mathbf{v}\times\mathbf{H}-\frac{1}{c^2}\mathbf{v}(\mathbf{v}\cdot\mathbf{E})\right\}.$$

§45. Gauge invariance

Let us consider to what extent the potentials are uniquely determined. The field is characterised by the effect which it produces on the motion of a charge located in it. But in the equation of motion (44.5) there appear not the potentials, but the field intensities **E** and **H**. Therefore two fields are physically identical if they are characterised by the same vectors **E** and **H**.

If we are given potentials **A** and ϕ, then these uniquely determine (according to (44.3) and (44.4)) the fields **E** and **H**. However, to one and the same field there can correspond different potentials. To show this, let us add to each component of the four-potential the quantity $-\partial f/\partial x^\mu$, where f is an arbitrary function of the coordinates and the time. Then the potential A_μ goes over into

$$A'_\mu = A_\mu - \frac{\partial f}{\partial x^\mu}. \tag{45.1}$$

As a result of this change there appears in the action integral (43.1) the additional term

$$\frac{e}{c}\frac{\partial f}{\partial x^{\mu}}\,\mathrm{d}x^{\mu} = \mathrm{d}\left(\frac{e}{c}f\right), \tag{45.2}$$

which is a total differential and has no effect on the equations of motion (cf. §2).

If in place of the four-potential we introduce the scalar and vector potentials, and in place of x^{μ}, the coordinates ct, x, y, z, then the four equations (45.1) can be written in the form

$$\mathbf{A}' = \mathbf{A} + \operatorname{grad} f, \quad \phi' = \phi - \frac{1}{c}\frac{\partial f}{\partial t}. \tag{45.3}$$

It is easy to check that electric and magnetic fields determined from equations (44.3) and (44.4) actually do not change upon replacement of $\mathbf{A}$ and ϕ by $\mathbf{A}'$ and ϕ'. Thus the transformation (45.3) does not change the fields. The potentials are therefore not uniquely defined; the vector potential is determined to within the gradient of an arbitrary function, and the scalar potential to within the time derivative of the same function.

In particular, we can add an arbitrary constant vector to the vector potential, and an arbitrary constant to the scalar potential. This is also clear directly from the fact that the definitions of $\mathbf{E}$ and $\mathbf{H}$ contain only derivatives of $\mathbf{A}$ and ϕ, and therefore the addition of constants to the latter does not affect the field intensities.

Only those quantities have physical meaning which are invariant with respect to the transformation (45.3) of the potentials; in particular all equations must be invariant under this transformation. This invariance is called *gauge invariance*.†

This non-uniqueness of the potentials gives us the possibility of choosing them so that they fulfil one auxiliary condition chosen by us. We emphasise that we can set one condition, since we may choose the

† We emphasise that this is related to the assumed constancy of e in (45.2). Thus the gauge invariance of the equations of electrodynamics and the conservation of charge are closely related to one another.

function f in (45.3) arbitrarily. In particular, it is always possible to choose the potentials so that ϕ is zero. If the vector potential is not zero, it is not generally possible to make it zero, since the condition $\mathbf{A} = 0$ represents three auxiliary conditions (for the three components of **A**).

§46. Constant electromagnetic field

By a constant electromagnetic field we mean a field which does not depend on the time. Clearly the potentials of a constant field can be chosen so that they are functions only of the coordinates and not of the time. A constant magnetic field is equal, as before, to $\mathbf{H} = \text{curl } \mathbf{A}$. A constant electric field is equal to

$$\mathbf{E} = -\text{grad } \phi. \tag{46.1}$$

Thus a constant electric field is determined only by the scalar potential and a constant magnetic field only by the vector potential.

We saw in the preceding section that the potentials are not uniquely determined. However, if we describe the constant electromagnetic field in terms of potentials which do not depend on the time, then we can add to the scalar potential, without changing the fields, only an arbitrary constant (not depending on either the coordinates or the time). Usually ϕ is subjected to the additional requirement that it be zero at infinity. Thus the arbitrary constant previously mentioned is determined, and the scalar potential of the constant field is thus determined uniquely. On the other hand, the vector potential is not uniquely determined: we can add to it the gradient of an arbitrary function of the coordinates.

We now determine the energy of a charge in a constant electromagnetic field. If the field is constant, then the Lagrangian for the charge also does not depend explicitly on the time. In this case the energy is conserved and coincides with the Hamiltonian. According to (43.6), we have

$$\mathcal{E} = \frac{mc^2}{\sqrt{[1-(v^2/c^2)]}} + e\phi. \tag{46.2}$$

Thus the presence of the field adds to the energy of the particle the term $e\phi$, the potential energy of the charge in the field. We note the important fact that the energy depends only on the scalar and not on the vector potential. This means that the magnetic field does not affect the energy of the charge. Only the electric field can change the energy of the particle. This is related to the fact already mentioned that the magnetic field, unlike the electric field, does no work on the charge.

If the field intensities are the same at all points in space, then the field is said to be *uniform*. The scalar potential of a uniform electric field can be expressed in terms of the field intensity as

$$\phi = -\mathbf{E}\cdot\mathbf{r}; \tag{46.3}$$

since $\mathbf{E}$ = constant,

$$\nabla(\mathbf{E}\cdot\mathbf{r}) = (\mathbf{E}\cdot\nabla)\mathbf{r} = \mathbf{E}.$$

The vector potential of a uniform magnetic field can be expressed in terms of its field intensity as

$$\mathbf{A} = \tfrac{1}{2}\mathbf{H}\times\mathbf{r}; \tag{46.4}$$

for $\mathbf{H}$ = constant, we obtain with the aid of well-known formulae of vector analysis

$$\text{curl}\,(\mathbf{H}\times\mathbf{r}) = \mathbf{H}\,\text{div}\,\mathbf{r} - (\mathbf{H}\cdot\nabla)\mathbf{r} = 2\mathbf{H}$$

(noting that div $\mathbf{r} = 3$).

§47. Motion in a constant uniform electric field

Let us consider the motion of a charge e in a uniform constant electric field $\mathbf{E}$. We take the direction of the field as the x axis. The motion will obviously proceed in a plane, which we choose as the xy plane. Then the equations of motion (44.5) become

$$\dot{p}_x = eE, \quad \dot{p}_y = 0,$$

so that

$$p_x = eEt, \quad p_y = p_0. \tag{47.1}$$

The time reference point has been chosen at the moment when $p_x = 0$; p_0 is the momentum of the particle at that moment.

The kinetic energy of the particle (the energy omitting the potential energy in the field) is $\mathscr{E}_{\text{kin}} = c\sqrt{(m^2c^2+p^2)}$. Substituting (47.1), we find in our case

$$\mathscr{E}_{\text{kin}} = \sqrt{[m^2c^4+c^2p_0^2+(ceEt)^2]} = \sqrt{[\mathscr{E}_0^2+(ceEt)^2]}, \tag{47.2}$$

where $\mathscr{E}_0$ is the energy at $t = 0$.

According to (39.11) the velocity of the particle is $\mathbf{v} = \mathbf{p}c^2/\mathscr{E}_{\text{kin}}$. For the velocity $v_x = \dot{x}$ we have therefore

$$\frac{\mathrm{d}x}{\mathrm{d}t} = \frac{p_x c^2}{\mathscr{E}_{\text{kin}}} = \frac{c^2 eEt}{\sqrt{[\mathscr{E}_0^2+(ceEt)^2]}}.$$

Integrating, we find

$$x = \frac{1}{eE}\sqrt{[\mathscr{E}_0^2+(ceEt)^2]}. \tag{47.3}$$

The constant of integration we set equal to zero.

For determining y, we have

$$\frac{\mathrm{d}y}{\mathrm{d}t} = \frac{p_y c^2}{\mathscr{E}_{\text{kin}}} = \frac{p_0 c^2}{\sqrt{[\mathscr{E}_0^2+(ceEt)^2]}},$$

from which

$$y = \frac{p_0 c}{eE}\sinh^{-1}\left(\frac{ceEt}{\mathscr{E}_0}\right). \tag{47.4}$$

We obtain the equation of the path by expressing t in terms of y from (47.4) and substituting in (47.3). This gives

$$x = \frac{\mathscr{E}_0}{eE}\cosh\frac{eEy}{p_0 c}. \tag{47.5}$$

Thus in a uniform electric field a charge moves along a catenary curve.

If the velocity of the particle is $v \ll c$, then we can take $p_0 = mv_0$, $\mathscr{E}_0 = mc^2$, and expand (47.5) in series in powers of $1/c$. Then we get, to within terms of higher order,

$$x = \frac{eE}{2mv_0^2}\, y^2 + \text{constant},$$

that is, the charge moves along a parabola, a result well known from classical mechanics.

§48. Motion in a constant uniform magnetic field

We now consider the motion of a charge e in a uniform magnetic field **H**. We choose the direction of the field as the z axis. We rewrite the equation of motion

$$\dot{\mathbf{p}} = \frac{e}{c}\,\mathbf{v}\times\mathbf{H}$$

in another form, by substituting for the momentum

$$\mathbf{p} = \frac{\mathscr{E}\mathbf{v}}{c^2},$$

where $\mathscr{E}$ is the energy of the particle, which is constant in the magnetic field. The equation of motion then goes over into the form

$$\frac{\mathscr{E}}{c^2}\frac{d\mathbf{v}}{dt} = \frac{e}{c}\,\mathbf{v}\times\mathbf{H} \tag{48.1}$$

or, expressed in terms of components,

$$\dot{v}_x = \omega v_y, \qquad \dot{v}_y = -\omega v_x, \qquad \dot{v}_z = 0, \tag{48.2}$$

where we have introduced the notation

$$\omega = \frac{ecH}{\mathscr{E}}. \tag{48.3}$$

We multiply the second equation (48.2) by i, and add it to the first:

$$\frac{d}{dt}(v_x + iv_y) = -i\omega(v_x + iv_y),$$

so that

$$v_x + iv_y = ae^{-i\omega t},$$

where a is a complex constant. This can be written in the form $a = v_{0t}e^{-i\alpha}$, where v_{0t} and α are real. Then

$$v_x + iv_y = v_{0t}e^{-i(\omega t+\alpha)}$$

and, separating real and imaginary parts, we find

$$v_x = v_{0t}\cos(\omega t+\alpha), \quad v_y = -v_{0t}\sin(\omega t+\alpha). \tag{48.4}$$

The constants v_{0t} and α are determined by the initial conditions; α is the initial phase, and from (48.4) it is clear that

$$v_{0t} = \sqrt{(v_x^2+v_y^2)},$$

that is, v_{0t} is the velocity of the particle in the xy plane, and stays constant throughout the motion.

From (48.4) we find, integrating once more,

$$x = x_0 + r\sin(\omega t+\alpha), \quad y = y_0 + r\cos(\omega t+\alpha), \tag{48.5}$$

where

$$r = \frac{v_{0t}}{\omega} = \frac{v_{0t}\mathcal{E}}{ecH} = \frac{cp_t}{eH} \tag{48.6}$$

(p_t is the projection of the momentum on the xy plane). From the third equation (48.2), we find $v_z = v_{0z}$ and

$$z = z_0 + v_{0z}t. \tag{48.7}$$

From (48.5) and (48.7), it is clear that the charge moves in a uniform magnetic field along a helix having its axis along the direction of the magnetic field and with a radius r given by (48.6). The velocity of the particle is constant. In the special case where $v_{0z} = 0$, that is, the charge has no velocity component along the field, it moves along a circle in a plane perpendicular to the field.

The quantity ω, as we see from the formulae, is the angular frequency of rotation of the particle in the plane perpendicular to the field.

If the velocity of the particle is low, then we can approximately put $\mathcal{E} = mc^2$. Then the frequency ω is changed to

$$\omega = \frac{eH}{mc}. \tag{48.8}$$

PROBLEMS

PROBLEM 1. Find the adiabatic invariant for the motion of a charge in a uniform magnetic field whose magnitude and direction vary slowly with time.

SOLUTION. Since the motion in the plane perpendicular to the magnetic field is periodic, the adiabatic invariant is the integral (cf. §32)

$$I = \frac{1}{2\pi} \oint \mathbf{P}_t \cdot d\mathbf{l},$$

taken over a complete period of the motion, i.e. over the circumference of a circle in the present case ($\mathbf{P}_t$ is the projection of the generalised momentum on the plane perpendicular to $\mathbf{H}$). Substituting $\mathbf{P}_t = \mathbf{p}_t + (e/c)\mathbf{A}$, we have

$$I = \frac{1}{2\pi} \oint \mathbf{p}_t \cdot d\mathbf{l} + \frac{e}{2\pi c} \oint \mathbf{A} \cdot d\mathbf{l}.$$

In the first term we note that $\mathbf{p}_t$ is constant in magnitude and directed along $d\mathbf{l}$; we apply Stokes' theorem to the second term and write curl $\mathbf{A} = \mathbf{H}$:

$$I = rp_t + \frac{e}{2c} Hr^2,$$

where r is the radius of the orbit. Substituting the expression (48.6) for r, we find

$$I = \frac{3cp_t^2}{2eH}.$$

From this we see that, for slow variation of H, the transverse momentum p_t varies proportionally to $\sqrt{H}$.

PROBLEM 2. Determine the frequencies of vibration of a charged space oscillator, placed in a constant, uniform magnetic field; the eigenfrequency of vibration of the oscillator (in the absence of the field) is ω_0.

SOLUTION. The equations of forced vibration of the oscillator in a magnetic field (directed along the z axis) are

$$\ddot{x} + \omega_0^2 x = \frac{eH}{mc}\dot{y}, \quad \ddot{y} + \omega_0^2 y = -\frac{eH}{mc}\dot{x}, \quad \ddot{z} + \omega_0^2 z = 0.$$

Multiplying the second equation by i and adding to the first, we find

$$\ddot{\xi}+\omega_0^2\xi = -i\frac{eH}{mc}\dot{\xi},$$

where $\xi = x+iy$. From this we find that the frequencies of vibration of the oscillator in a plane perpendicular to the field are

$$\omega = \sqrt{\left[\omega_0^2+\frac{1}{4}\left(\frac{eH}{mc}\right)^2\right]}\pm\frac{eH}{2mc}.$$

If the field H is weak, this formula goes over into

$$\omega = \omega_0 \pm eH/2mc.$$

The vibration along the direction of the field remains unchanged.

§49. Motion of a charge in crossed fields

Finally we consider the motion of a charge in the case where there are present both electric and magnetic fields, constant and uniform. We limit ourselves to the non-relativistic case, where the velocity of the charge $v \ll c$, so that its momentum $\mathbf{p} = m\mathbf{v}$; as we shall see later, it is necessary for this that the electric field be small compared with the magnetic field.

We choose the direction of $\mathbf{H}$ as the z axis, and the plane passing through $\mathbf{H}$ and $\mathbf{E}$ as the yz plane. Then the equation of motion

$$m\dot{\mathbf{v}} = e\mathbf{E}+\frac{e}{c}\mathbf{v}\times\mathbf{H}$$

can be written in the form

$$m\ddot{x} = \frac{e}{c}\dot{y}H, \quad m\ddot{y} = eE_y - \frac{e}{c}\dot{x}H, \quad m\ddot{z} = eE_z. \tag{49.1}$$

From the third equation we see that the charge moves with uniform acceleration in the z direction, that is,

$$z = \frac{eE_z}{2m}t^2 + v_{0z}t. \tag{49.2}$$

Multiplying the second equation (49.1) by i and adding to the first, we find

$$\frac{\mathrm{d}}{\mathrm{d}t}(\dot{x}+i\dot{y})+i\omega(\dot{x}+i\dot{y}) = i\frac{e}{m}E_y$$

($\omega = eH/mc$). The integral of this equation is equal to the sum of the integral of the same equation without the right-hand term and a particular integral of the equation with the right-hand term. The first of these is $ae^{-i\omega t}$, the second is $eE_y/m\omega = cE_y/H$. Thus

$$\dot{x}+i\dot{y} = ae^{-i\omega t}+cE_y/H.$$

The constant a is in general complex. Writing it in the form $a = be^{i\alpha}$, with real b and α, we see that, since a is multiplied by $e^{-i\omega t}$, we can, by a suitable choice of the time origin, give the phase α any arbitrary value. We choose this so that a is real. Then breaking up $\dot{x}+i\dot{y}$ into real and imaginary parts, we find

$$\dot{x} = a\cos\omega t+cE_y/H, \quad \dot{y} = -a\sin\omega t. \tag{49.3}$$

At $t = 0$ the velocity is along the x axis. We see that the components of the velocity of the particle are periodic functions of the time; their average values are

$$\bar{v}_x = cE_y/H, \quad \bar{v}_y = 0. \tag{49.4}$$

This average velocity of a charge in crossed electric and magnetic fields is often called the *electromagnetic drift* velocity. Its direction is perpendicular to both fields and is independent of the sign of the charge.

All the formulae of this section assume that the velocity of the particle is small compared with the velocity of light; we see that, for this to be so, it is necessary in particular that the electric and magnetic fields satisfy the condition

$$E_y/H \ll 1, \tag{49.5}$$

while the absolute magnitudes of E_y and H can be arbitrary.

Integrating equation (49.3) again, and choosing the constant of integration so that at $t = 0$, $x = y = 0$, we obtain

$$x = \frac{a}{\omega}\sin\omega t + \frac{cE_y}{H}t; \quad y = \frac{a}{\omega}(\cos\omega t - 1). \tag{49.6}$$

Considered as parametric equations of a curve, these equations define a trochoid. Depending on whether a is larger or smaller in

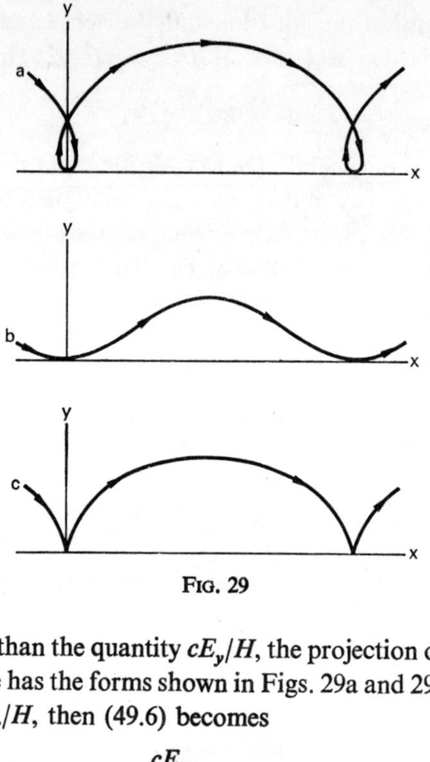

FIG. 29

absolute value than the quantity cE_y/H, the projection of the trajectory on the xy plane has the forms shown in Figs. 29a and 29b, respectively.

If $a = -cE_y/H$, then (49.6) becomes

$$x = \frac{cE_y}{\omega H}(\omega t - \sin\omega t),$$
$$y = \frac{cE_y}{\omega H}(1 - \cos\omega t); \tag{49.7}$$

that is, the projection of the trajectory on the xy plane is a cycloid (Fig. 29c).

§50. The electromagnetic field tensor

Formulae (44.3), (44.4), expressing the field intensities in terms of the potentials, are written in three-dimensional form and are therefore not suitable for getting the law of transformation of these quantities under a change of reference frame.

It is easy to see that the set of components of the two three-dimensional vectors **E** and **H** can be described as the components of an antisymmetric four-tensor:

$$F_{\mu\nu} = \frac{\partial A_\nu}{\partial x^\mu} - \frac{\partial A_\mu}{\partial x^\nu} \tag{50.1}$$

(called the *electromagnetic field tensor*). The meaning of the individual components can be seen by substituting the values $A_\mu = (\phi, -\mathbf{A})$ in the definition (50.1). The result can be written in the form of a matrix in which the index $\mu = 0, 1, 2, 3$ labels the rows, and the index ν the columns:

$$F_{\mu\nu} = \begin{pmatrix} 0 & E_x & E_y & E_z \\ -E_x & 0 & -H_z & H_y \\ -E_y & H_z & 0 & -H_x \\ -E_z & -H_y & H_x & 0 \end{pmatrix}. \tag{50.2}$$

The contravariant components of the same tensor have a sign change whenever a spatial index is raised:

$$F^{\mu\nu} = \begin{pmatrix} 0 & -E_x & -E_y & -E_z \\ E_x & 0 & -H_z & H_y \\ E_y & H_z & 0 & -H_x \\ E_z & -H_y & H_x & 0 \end{pmatrix}. \tag{50.3}$$

We note that the equations of motion of a charge in the field can be written, using the tensor $F_{\mu\nu}$, in the form

$$\frac{\mathrm{d}p^\mu}{\mathrm{d}s} = \frac{e}{c} F^{\mu\nu} u_\nu . \tag{50.4}$$

Writing out the expressions on both sides of the equation using the three-dimensional notation (40.2), (40.5), and (50.3) (and substituting $ds = c\,dt\sqrt{[1-(v^2/c^2)]}$, we see that for $\mu = 1, 2, 3$ we get the three components of the vector equation (44.6), while for $\mu = 0$ we find the work equation (44.7).

The transformation formulae for the fields **E** and **H** can now be found using the general rules for transformation of four-tensors. The components of the four-tensor of rank two $F^{\mu\nu}$ transform like a product of coordinates $x^\mu x^\nu$. Under the Lorentz transformation (36.3) the coordinates $x^2 = y$ and $x^3 = z$ do not change, so the component F^{23} does not change:

$$F^{23} = F'^{23}.$$

For the same reason, the components F^{02}, F^{03} and F^{12}, F^{13} transform respectively like the coordinates $x^0 = ct$ and $x^1 = x$:

$$F^{02} = \frac{F'^{02}+(VF'^{12}/c)}{\sqrt{[1-(V^2/c^2)]}}, \quad F^{12} = \frac{F'^{12}+(VF'^{02}/c)}{\sqrt{[1-(V^2/c^2)]}}$$

and similarly for F^{03}, F^{13}. Finally, the component F^{01} should transform like the product x^0x^1; one would then get

$$F^{01} = \frac{1}{1-(V^2/c^2)}\left\{F'^{01}+\frac{V^2}{c^2}F'^{10}+\frac{V}{c}(F'^{01}+F'^{10})\right\}.$$

But since in the present case the tensor $F^{\mu\nu}$ is antisymmetric, $F'^{01} = -F'^{10}$ and so

$$F^{01} = F'^{01}.$$

Now expressing the components of the tensor $F^{\mu\nu}$ in terms of the components of the fields **E** and **H** according to (50.3), we find the following transformation formulae for the electric field:

$$E_x = E'_x, \quad E_y = \frac{E'_y+(VH'_z/c)}{\sqrt{[1-(V^2/c^2)]}}, \quad E_z = \frac{E'_z-(VH'_y/c)}{\sqrt{[1-(V^2/c^2)]}}, \tag{50.5}$$

and for the magnetic field:

$$H_x = H'_x, \quad H_y = \frac{H'_y-(VE'_z/c)}{\sqrt{[1-(V^2/c^2)]}}, \quad H_z = \frac{H'_z+(VE'_y/c)}{\sqrt{[1-(V^2/c^2)]}}. \tag{50.6}$$

Thus the electric and magnetic fields, like the majority of physical quantities, are relative; that is, their properties are different in different reference frames. In particular, the electric or the magnetic field can be equal to zero in one reference frame and at the same time be present in another frame.

If the magnetic field $\mathbf{H}' = 0$ in the frame K', then, on the basis of (50.5) and (50.6), the following relation exists between the electric and magnetic fields in the frame K:

$$\mathbf{H} = \frac{1}{c}\,\mathbf{V}\times\mathbf{E}. \tag{50.7}$$

If, in the frame K', $\mathbf{E}' = 0$, then in the frame K

$$\mathbf{E} = -\frac{1}{c}\,\mathbf{V}\times\mathbf{H}. \tag{50.8}$$

Consequently, in both cases, in the frame K the magnetic and electric fields are mutually perpendicular. These formulae also have a significance when used in the reverse direction: if the fields $\mathbf{E}$ and $\mathbf{H}$ are mutually perpendicular (but not equal in magnitude) in some reference frame K, then there exists a reference frame K' in which the field is purely electric or purely magnetic.

§51. Invariants of the field

From the electric and magnetic field intensities we can form invariant quantities, which remain unchanged in the transition from one inertial frame of reference to another.

We obtain such a quantity by forming the four-dimensional scalar $F_{\mu\nu}F^{\mu\nu}$. Writing it out in three-dimensional notation, we find that $F_{\mu\nu}F^{\mu\nu} = 2(H^2-E^2)$. Thus one of the required invariants is the quantity

$$H^2-E^2 = \text{invariant}. \tag{51.1}$$

By a direct check using formulae (50.5), (50.6) one easily verifies that the sum $E_xH_x+E_yH_y+E_zH_z$ is left unchanged by a Lorentz transformation Thus

$$\mathbf{E}\cdot\mathbf{H} = \text{invariant.} \tag{51.2}$$

There is, however, a fundamental difference between these two invariants with respect to their behaviour under reflection (*inversion*) of the spatial coordinate system (simultaneous change of sign of x, y, z). We recall that under such a transformation the components of a true (*polar*) vector change sign. But the components of a vector that can be written as the vector product of two polar vectors remain unchanged under inversion (such vectors are said to be *axial*). The scalar product of two axial or two polar vectors is a true scalar—it is not changed by inversion. The scalar product of an axial and a polar vector is a *pseudoscalar*—it changes sign under inversion.

But according to the definition (44.3), (44.4), **E** is a polar vector while **H** is an axial vector (the vector product of the polar vectors ∇ and **A**). It is then clear that H^2-E^2 is a true scalar, while $\mathbf{E}\cdot\mathbf{H}$ is a pseudoscalar (while its square $(\mathbf{E}\cdot\mathbf{H})^2$ will be a true scalar).

We note some consequences of the invariance of the expressions (51.1), (51.2). If in some frame of reference the fields **E** and **H** are equal in magnitude ($E^2 = H^2$), then they are equal in magnitude in every other inertial reference frame. If the fields **E** and **H** are perpendicular in some reference frame ($\mathbf{E}\cdot\mathbf{H} = 0$), then they are also perpendicular in every other frame.

If $\mathbf{E}\cdot\mathbf{H} = 0$, we can find a frame of reference in which either $\mathbf{E} = 0$ or $\mathbf{H} = 0$ (depending on whether $E^2-H^2 < 0$ or > 0), i.e. the field is purely magnetic or purely electric. Conversely, if in some reference frame $\mathbf{E} = 0$ or $\mathbf{H} = 0$, then they will be perpendicular in every other frame, in accordance with our remarks at the end of the preceding section.

An exceptional case occurs when both invariants are zero: $\mathbf{E}\cdot\mathbf{H} = 0$ and $E^2-H^2 = 0$. In this case **E** and **H** are equal in magnitude and mutually perpendicular in every frame of reference.

CHAPTER 11

THE ELECTROMAGNETIC FIELD EQUATIONS

§52. The first pair of Maxwell's equations

From the expressions

$$\mathbf{H} = \operatorname{curl} \mathbf{A}, \quad \mathbf{E} = -\frac{1}{c}\frac{\partial \mathbf{A}}{\partial t} - \operatorname{grad} \phi$$

it is easy to obtain equations containing only **E** and **H**. To do this we find curl **E**:

$$\operatorname{curl} \mathbf{E} = -\frac{1}{c}\frac{\partial}{\partial t} \operatorname{curl} \mathbf{A} - \operatorname{curl} \operatorname{grad} \phi.$$

But the curl of any gradient is zero. Consequently,

$$\operatorname{curl} \mathbf{E} = -\frac{1}{c}\frac{\partial \mathbf{H}}{\partial t}. \qquad (52.1)$$

Taking the divergence of both sides of the equation curl **A** = **H**, and recalling that div curl = 0, we find

$$\operatorname{div} \mathbf{H} = 0. \qquad (52.2)$$

The equations (52.1) and (52.2) are called the first pair of *Maxwell's equations*.† We note that these two equations still do not completely

† Maxwell's equations (the fundamental equations of electrodynamics) were first formulated by him in the 1860's.

determine the properties of the field. This is clear from the fact that they determine the change of the magnetic field with time (the derivative $\partial \mathbf{H}/\partial t$), but do not determine the derivative $\partial \mathbf{E}/\partial t$.

Equations (52.1) and (52.2) can be written in integral form. According to Gauss' theorem

$$\int \operatorname{div} \mathbf{H}\, dV = \oint \mathbf{H} \cdot d\mathbf{f},$$

where the integral on the right goes over the entire closed surface surrounding the volume over which the integral on the left is extended. On the basis of (52.2), we have

$$\oint \mathbf{H} \cdot d\mathbf{f} = 0. \tag{52.3}$$

The integral of a vector over a surface is called the *flux of the vector* through the surface. Thus the flux of the magnetic field through every closed surface is zero.

According to Stokes' theorem,

$$\int \operatorname{curl} \mathbf{E} \cdot d\mathbf{f} = \oint \mathbf{E} \cdot d\mathbf{l},$$

where the integral on the right is taken over the closed contour bounding the surface over which the integral on the left is taken. From (52.1) we find, integrating both sides over any surface,

$$\oint \mathbf{E} \cdot d\mathbf{l} = -\frac{1}{c}\frac{\partial}{\partial t}\int \mathbf{H} \cdot d\mathbf{f}. \tag{52.4}$$

The integral of a vector over a closed contour is called the *circulation* of the vector around the contour. The circulation of the electric field is also called the *electromotive force* in the given contour. Thus the electromotive force in any contour is equal to minus the time derivative of the magnetic flux through a surface bounded by this contour.

§53. The action function of the electromagnetic field

The action function S for the whole system, consisting of an electromagnetic field as well as the particles located in it, must consist of three parts:

$$S = S_f + S_m + S_{mf}. \tag{53.1}$$

S_m is that part of the action which depends only on the properties of the particles, that is, just the action for free particles. For a single free particle, it is given by (39.1). If there are several particles, then their total action is the sum of the actions for each of the individual particles. Thus,

$$S_m = -\sum mc \int ds. \tag{53.2}$$

The quantity S_{mf} is that part of the action which depends on the interaction between the particles and the field. According to §43, we have for a system of particles

$$S_{mf} = -\sum \frac{e}{c} \int A_\mu \, dx^\mu. \tag{53.3}$$

In each term of this sum, A_μ is the potential of the field at that point of space-time at which the corresponding particle is located. The sum S_m+S_{mf} is already familiar to us as the action (43.1) for charges in a field.

Finally S_f is that part of the action which depends only on the properties of the field itself, that is, S_f is the action for a field in the absence of charges. Up to now, because we were interested only in the motion of charges in a *given* electromagnetic field, the quantity S_f, which does not depend on the particles, did not concern us, since this term cannot affect the motion of the particles. Nevertheless this term is necessary when we want to find equations determining the field itself. This corresponds to the fact that from the parts S_m+S_{mf} of the action we found only two equations for the field, (52.1) and (52.2), which are not sufficient for complete determination of the field.

To establish the form of the action S_f for the field, we start from the following very important property of electromagnetic fields. As experiment shows, the electromagnetic field satisfies the *principle of superposition.* This principle consists in the statement that the field produced by a system of charges is the result of a simple composition of the fields produced by each of the charges individually. This means that the resultant field intensity at each point is equal to the vector sum of the individual field intensities at that point.

Every solution of the field equations gives a field that can exist in nature. According to the principle of superposition, the sum of any such fields must be a field that can exist in nature, that is, must satisfy the field equations.

As is well known, linear differential equations have just this property, that the sum of any solutions is also a solution. Consequently the field equations must be linear differential equations.

From the above discussion, it follows that under the integral sign in the expression for the action S_f there must stand an expression quadratic in the field. Only in this case will the field equations be linear; the field equations are obtained by varying the action, and in the variation the degree of the integrand decreases by unity.

The potentials cannot enter into the expression for the action S_f, since they are not uniquely determined (in S_{mf} this lack of uniqueness was not important). Therefore S_f must be the integral of some function of the electromagnetic field tensor $F_{\mu\nu}$. But the action must be a scalar and must therefore be the integral of some (true) scalar. The only such quantity is the product $F_{\mu\nu}F^{\mu\nu}$.†

Thus S_f must have the form

$$S_f = a \iint F_{\mu\nu}F^{\mu\nu}\, \mathrm{d}V\, \mathrm{d}t, \qquad \mathrm{d}V = \mathrm{d}x\, \mathrm{d}y\, \mathrm{d}z,$$

where the integral extends over all space and over the time between two given moments; a is some constant. Under the integral stands $F_{\mu\nu}F^{\mu\nu} = 2(H^2 - E^2)$. The field $\mathbf{E}$ contains the derivative $\partial\mathbf{A}/\partial t$; but it is easy to see that $(\partial\mathbf{A}/\partial t)^2$ must appear in the action with the positive sign (and therefore E^2 must have a positive sign). For if $(\partial\mathbf{A}/\partial t)^2$ appeared in S_f with a minus sign, then sufficiently rapid change of the potential with time (in the time interval under consideration) could always make S_f a negative quantity with arbitrarily large absolute

† The function in the integrand of S_f must not include derivatives of $F_{\mu\nu}$, since the Lagrangian can contain, aside from the coordinates, only their first time derivatives; the role of "coordinates" (i.e., parameters to be varied in the principle of least action) is in this case played by the field potentials A_μ; this is analogous to the situation in mechanics where the Lagrangian of a mechanical system contains only the coordinates of the particles and their first time derivatives.

value. Consequently S_f could not have a minimum, as is required by the principle of least action. Thus, a must be negative.

The numerical value of a depends on the choice of units for measurement of the field. We note that, after the choice of a definite value for a and for the units of measurement of the field, the units for measurement of all other electromagnetic quantities are determined. From now on we shall use the *Gaussian system of units*; in this system a is a dimensionless quantity, equal to $-(1/16\pi)$.

Thus the action for the field has the form

$$S_f = -\frac{1}{16\pi c}\int F_{\mu\nu}F^{\mu\nu}\,d\Omega, \qquad d\Omega = c\,dt\,dx\,dy\,dz. \tag{53.4}$$

In three-dimensional form:

$$S_f = \frac{1}{8\pi}\iint (E^2 - H^2)\,dV\,dt. \tag{53.5}$$

In other words, the Lagrangian for the field is

$$L_f = \frac{1}{8\pi}\int (E^2 - H^2)\,dV. \tag{53.6}$$

The action for field plus charges has the form

$$S_f = -\sum\int mc\,ds - \sum\int \frac{e}{c}A_\mu\,dx^\mu - \frac{1}{16\pi c}\int F_{\mu\nu}F^{\mu\nu}\,d\Omega. \tag{53.7}$$

We note that now the charges are no longer assumed to be small, as in the derivation of the equation of motion of a charge in a given field. Therefore A_μ and $F_{\mu\nu}$ refer to the actual field, that is, the external field plus the field produced by the charges themselves; A_μ and $F_{\mu\nu}$ now depend on the positions and velocities of the charges.

§54. The current four-vector

Instead of treating charges as points, for mathematical convenience we frequently consider them to be distributed continuously in space. Then we can introduce the *charge density* ϱ such that $\varrho\,dV$ is the charge contained in the volume $\mathrm{d}V$. The density ϱ is in general a function of the coordinates and the time. The integral of ϱ over a certain volume is the charge contained in that volume.

Here we must remember that charges are actually point-like, so that the density ϱ is zero everywhere except at points where charges are located, and the integral $\int \varrho\, \mathrm{d}V$ must be equal to the sum of the charges contained in the given volume. Therefore ϱ can be expressed with the help of the δ-function† in the following form:

$$\varrho = \sum_a e_a \delta(\mathbf{r}-\mathbf{r}_a) \tag{54.1}$$

where the sum goes over all the charges and $\mathbf{r}_a$ is the position vector of the charge e_a.

† The δ-function $\delta(x)$ is defined as follows: $\delta(x) = 0$, for all non-zero values of x; for $x = 0$, $\delta(0) = \infty$, in such a way that the integral

$$\int_{-\infty}^{+\infty} \delta(x)\, \mathrm{d}x = 1. \tag{I}$$

From this definition there result the following properties: if $f(x)$ is any continuous function, then

$$\int_{-\infty}^{+\infty} f(x)\, \delta(x-a)\, \mathrm{d}x = f(a), \tag{II}$$

and in particular,

$$\int_{-\infty}^{+\infty} f(x)\, \delta(x)\, \mathrm{d}x = f(0). \tag{III}$$

(The limits of integration, of course, need not be $\pm\infty$; the range of integration can be arbitrary, provided that it includes the point at which the δ-function does not vanish.)

We also have

$$\delta(-x) = \delta(x), \qquad \delta(ax) = \frac{1}{|a|}\, \delta(x), \tag{IV}$$

in the sense that both sides give the same result when they appear as a factor in an integrand.

In the same way as $\delta(x)$ was defined for one variable x, we can introduce a three-dimensional δ-function, $\delta(\mathbf{r})$, equal to zero everywhere except at the origin of the three-dimensional coordinate system, and whose integral over all space is unity. As such a function we can use the product $\delta(x)\, \delta(y)\, \delta(z)$.

The charge of a particle is, from its very definition, an invariant quantity, that is, it does not depend on the choice of reference frame. On the other hand, the density ϱ is not an invariant—only the product $\varrho\,dV$ is invariant.

Multiplying the equality $de = \varrho\,dV$ on both sides by dx^μ, we have

$$de\,dx^\mu = \varrho\,dV\,dx^\mu = \varrho\,dV\,dt\,\frac{dx^\mu}{dt}.$$

On the left stands a four-vector (since de is a scalar and dx^μ is a four-vector). This means that the right-hand side must be a four-vector. But $dV\,dt$ is a scalar, and so $\varrho(dx^\mu/dt)$ is a four-vector. This vector (we denote it by j^μ) is called the *current four-vector*:

$$j^\mu = \varrho\frac{dx^\mu}{dt}. \tag{54.2}$$

The three space components of this vector form a vector in ordinary space,

$$\mathbf{j} = \varrho\mathbf{v}, \tag{54.3}$$

where $\mathbf{v}$ is the velocity of the charge at the given point. The vector $\mathbf{j}$ is called the *current density vector*. The time component of the current four-vector is $c\varrho$. Thus

$$j^\mu = (c\varrho, \mathbf{j}). \tag{54.4}$$

Let us introduce the current four-vector into the expression (53.7) for the action and transform the second term in that expression. Introducing in place of the point charges e a continuous distribution of charge with density ϱ, we must write this term as

$$-\frac{1}{c}\int \varrho A_\mu\,dx^\mu\,dV,$$

replacing the sum over the charges by an integral over the whole volume. We rewrite it in the form

$$-\frac{1}{c}\int \varrho\frac{dx^\mu}{dt}A_\mu\,dV\,dt = -\frac{1}{c^2}\int A_\mu j^\mu\,d\Omega.$$

Thus the total action S takes the form

$$S = -\sum \int mc\,\mathrm{d}s - \frac{1}{c^2}\int A_\mu j^\mu\,\mathrm{d}\Omega - \frac{1}{16\pi c}\int F_{\mu\nu}F^{\mu\nu}\,\mathrm{d}\Omega. \quad (54.5)$$

§55. The equation of continuity

The change with time of the charge contained in a certain volume is determined by the derivative

$$\frac{\partial}{\partial t}\int \varrho\,\mathrm{d}V.$$

On the other hand, the change in unit time is determined by the quantity of charge which in unit time leaves the volume and goes to the outside or, conversely, passes to its interior. The quantity of charge which passes in unit time though the element $\mathbf{df}$ of the surface bounding the volume is equal to $\varrho\mathbf{v}\cdot\mathbf{df}$, where $\mathbf{v}$ is the velocity of the charge at the point in space where the element $\mathbf{df}$ is located. The vector $\mathbf{df}$ is directed, as always, along the external normal to the surface, that is, along the normal toward the outside of the volume under consideration. Therefore $\varrho\mathbf{v}\cdot\mathbf{df} \equiv \mathbf{j}\cdot\mathbf{df}$ is positive if charge leaves the volume, and negative if charge enters the volume. The total amount of charge leaving the given volume per unit time is consequently $\oint \mathbf{j}\cdot\mathbf{df}$, where the integral extends over the whole of the closed surface bounding the volume.

From the equality of these two expressions, we get

$$\frac{\partial}{\partial t}\int \varrho\,\mathrm{d}V = -\oint \mathbf{j}\cdot\mathbf{df}. \quad (55.1)$$

The minus sign appears on the right, since the left-hand side is positive if the total charge in the given volume increases. This equation is the *equation of continuity*, expressing the conservation of charge, in integral form.

We also write this equation in differential form. To do this we apply Gauss' theorem to the right-hand side of (55.1):

$$\oint \mathbf{j}\cdot\mathbf{df} = \int \operatorname{div}\mathbf{j}\,\mathrm{d}V,$$

and find

$$\int \left(\operatorname{div} \mathbf{j} + \frac{\partial \varrho}{\partial t}\right) \mathrm{d}V = 0.$$

Since this must hold for integration over an arbitrary volume, the integrand must be zero:

$$\operatorname{div} \mathbf{j} + \frac{\partial \varrho}{\partial t} = 0. \tag{55.2}$$

This is the equation of continuity in differential form.

It is easy to check that the expression (54.1) for ϱ in δ-function form necessarily satisfies the equation of continuity. For simplicity we assume that we have only one charge, so that

$$\varrho = e\delta(\mathbf{r} - \mathbf{r}_0).$$

The current $\mathbf{j}$ is then

$$\mathbf{j} = e\mathbf{v}\,\delta(\mathbf{r} - \mathbf{r}_0),$$

where $\mathbf{v}$ is the velocity of the charge. We determine the derivative $\partial\varrho/\partial t$. During the motion of the charge its coordinates change, that is, the vector $\mathbf{r}_0$ changes. Therefore

$$\frac{\partial \varrho}{\partial t} = \frac{\partial \varrho}{\partial \mathbf{r}_0} \cdot \frac{\partial \mathbf{r}_0}{\partial t}.$$

But $\partial\mathbf{r}_0/\partial t$ is just the velocity $\mathbf{v}$ of the charge. Furthermore, since ϱ is a function of $\mathbf{r} - \mathbf{r}_0$,

$$\frac{\partial \varrho}{\partial \mathbf{r}_0} = -\frac{\partial \varrho}{\partial \mathbf{r}}.$$

Consequently

$$\frac{\partial \varrho}{\partial t} = -\mathbf{v} \cdot \operatorname{grad} \varrho = -\operatorname{div}(\varrho \mathbf{v})$$

(the velocity $\mathbf{v}$ of the charge of course does not depend on $\mathbf{r}$). Thus we arrive at the equation (55.2).

In four-dimensional form the continuity equation (55.2) is expressed by the statement that the four-divergence of the current four-vector is zero:

$$\frac{\partial j^\mu}{\partial x^\mu} = 0. \tag{55.3}$$

§56. The second pair of Maxwell's equations

In finding the field equations from the principle of least action, we must assume the motion of the charges to be given and vary only the potentials of the field, which here play the role of "generalised coordinates" of the system. (In finding the equations of motion of a particle we did the opposite, assuming the field to be given and varying the path of the particle.) This derivation is conveniently done in four-dimensional form.

According to the above remarks the variation of the first term in (54.5) is now zero, and we must not vary the current j^μ in the second term. Thus,

$$\delta S = -\frac{1}{c}\int\left[\frac{1}{c}\, j^\mu \delta A_\mu + \frac{1}{8\pi}\, F^{\mu\nu}\delta F_{\mu\nu}\right] d\Omega = 0$$

(where we have used the fact that $F^{\mu\nu}\delta F_{\mu\nu} = F_{\mu\nu}\delta F^{\mu\nu}$ in varying the second term). Substituting

$$F_{\mu\nu} = \frac{\partial A_\nu}{\partial x^\mu} - \frac{\partial A_\mu}{\partial x^\nu}$$

in the factor $\delta F_{\mu\nu}$, we have

$$\delta S = -\frac{1}{c}\int\left\{\frac{1}{c}\, j^\mu \delta A_\mu + \frac{1}{8\pi}\left(F^{\mu\nu}\frac{\partial}{\partial x^\mu}\,\delta A_\nu - F^{\mu\nu}\frac{\partial}{\partial x^\nu}\,\delta A_\mu\right)\right\} d\Omega.$$

In the second term we interchange the dummy indices μ and ν, and also replace $F^{\nu\mu}$ by $-F^{\mu\nu}$. After this the second and third terms are identical, so that

$$\delta S = -\frac{1}{c}\int\left\{\frac{1}{c}\, j^\mu \delta A_\mu - \frac{1}{4\pi}\, F^{\mu\nu}\frac{\partial}{\partial x^\nu}\,\delta A_\mu\right\} d\Omega.$$

We also write

$$-\frac{1}{4\pi}\, F^{\mu\nu}\frac{\partial}{\partial x^\nu}\,\delta A_\mu = -\frac{1}{4\pi}\frac{\partial}{\partial x^\nu}\,(F^{\mu\nu}\delta A_\mu) + \frac{1}{4\pi}\,\delta A_\mu\,\frac{\partial F^{\mu\nu}}{\partial x^\nu}$$

and, applying the four-dimensional form of Gauss' theorem (38.9) to the integral of the first term, we find

$$\delta S = -\frac{1}{c}\int\left\{\frac{1}{c}j^{\mu}+\frac{1}{4\pi}\frac{\partial F^{\mu\nu}}{\partial x^{\nu}}\right\}\delta A_{\mu}\,\mathrm{d}\Omega-\frac{1}{4\pi c}\int F^{\mu\nu}\delta A_{\mu}\,\mathrm{d}S_{\nu}. \quad (56.1)$$

In the last term we must take the values at the limits of integration. The limits for the spatial coordinates are at infinity, where the field is zero. At the limits of the time integration, that is, at the given initial and final time values, the variation of the potentials is zero, since in accordance with the principle of least action the potentials are given at these times. Thus the last term in (56.1) is zero, and we find

$$\int\left(\frac{1}{c}j^{\mu}+\frac{1}{4\pi}\frac{\partial F^{\mu\nu}}{\partial x^{\nu}}\right)\delta A_{\mu}\,\mathrm{d}\Omega = 0.$$

Since the variations δA_{μ} are arbitrary, the quantity in parentheses must be zero:

$$\frac{\partial F^{\mu\nu}}{\partial x^{\nu}} = -\frac{4\pi}{c}j^{\mu}. \quad (56.2)$$

Let us express these four ($\mu = 0, 1, 2, 3$) equations in three-dimensional form. For $\mu = 1$,

$$\frac{\partial F^{11}}{\partial x}+\frac{\partial F^{12}}{\partial y}+\frac{\partial F^{13}}{\partial z}+\frac{1}{c}\frac{\partial F^{10}}{\partial t} = -\frac{4\pi}{c}j^{1}.$$

Substituting the values of the components of $F^{\mu\nu}$, we find

$$\frac{\partial H_z}{\partial y}-\frac{\partial H_y}{\partial z}-\frac{1}{c}\frac{\partial E_x}{\partial t} = \frac{4\pi}{c}j_x.$$

This together with the two succeeding equations ($\mu = 2, 3$) can be written as one vector equation:

$$\operatorname{curl}\mathbf{H} = \frac{1}{c}\frac{\partial \mathbf{E}}{\partial t}+\frac{4\pi}{c}\mathbf{j}. \quad (56.3)$$

Finally, the fourth equation ($\mu = 0$) gives

$$\operatorname{div}\mathbf{E} = 4\pi\varrho. \quad (56.4)$$

Equations (56.3) and (56.4) are the second pair of Maxwell's equations.[†] Together with the first pair of Maxwell's equations they completely determine the electromagnetic field, and are the fundamental equations of the theory of such fields, i.e. of *electrodynamics*.

Let us write these equations in integral form. Integrating (56.4) over a volume and applying Gauss' theorem

$$\int \operatorname{div} \mathbf{E}\, dV = \oint \mathbf{E}\cdot d\mathbf{f},$$

we get

$$\oint \mathbf{E}\cdot d\mathbf{f} = 4\pi \int \varrho\, dV. \tag{56.5}$$

Thus the flux of the electric field through a closed surface is equal to 4π times the total charge contained in the volume bounded by the surface.

Integrating (56.3) over an open surface and applying Stokes' theorem

$$\int \operatorname{curl} \mathbf{H}\cdot d\mathbf{f} = \oint \mathbf{H}\cdot d\mathbf{l},$$

we find

$$\oint \mathbf{H}\cdot d\mathbf{l} = \frac{1}{c}\frac{\partial}{\partial t}\int \mathbf{E}\cdot d\mathbf{f} + \frac{4\pi}{c}\int \mathbf{j}\cdot d\mathbf{f}. \tag{56.6}$$

The quantity

$$\frac{1}{4\pi}\frac{\partial \mathbf{E}}{\partial t} \tag{56.7}$$

is called the *displacement current*. From (56.6) written in the form

$$\oint \mathbf{H}\cdot d\mathbf{l} = \frac{4\pi}{c}\int \left(\mathbf{j} + \frac{1}{4\pi}\frac{\partial \mathbf{E}}{\partial t}\right)\cdot d\mathbf{f},$$

we see that the circulation of the magnetic field around any contour is equal to $4\pi/c$ times the sum of the true current and displacement current passing through a surface bounded by this contour.

---

† Maxwell's equations in a form applicable to point charges in the electromagnetic field in vacuum were formulated by Lorentz.

From Maxwell's equations we can obtain the already familiar continuity equation. Taking the divergence of both sides of (56.3), we find

$$\operatorname{div}\operatorname{curl}\mathbf{H} = \frac{1}{c}\frac{\partial}{\partial t}\operatorname{div}\mathbf{E} + \frac{4\pi}{c}\operatorname{div}\mathbf{j}.$$

But $\operatorname{div}\operatorname{curl}\mathbf{H} \equiv 0$ and $\operatorname{div}\mathbf{E} = 4\pi\varrho$, according to (56.4). Thus we arrive once more at equation (55.2).

§57. Energy density and energy flux

Let us multiply both sides of (56.3) by $\mathbf{E}$ and both sides of (52.1) by $\mathbf{H}$ and subtract the resultant equations. Then we get

$$\frac{1}{c}\mathbf{E}\cdot\frac{\partial \mathbf{E}}{\partial t} + \frac{1}{c}\mathbf{H}\cdot\frac{\partial \mathbf{H}}{\partial t} = -\frac{4\pi}{c}\mathbf{j}\cdot\mathbf{E} - (\mathbf{H}\cdot\operatorname{curl}\mathbf{E} - \mathbf{E}\cdot\operatorname{curl}\mathbf{H}).$$

Using the well-known formula of vector analysis,

$$\operatorname{div}(\mathbf{a}\times\mathbf{b}) = \mathbf{b}\cdot\operatorname{curl}\mathbf{a} - \mathbf{a}\cdot\operatorname{curl}\mathbf{b},$$

we rewrite this relation in the form

$$\frac{1}{2c}\frac{\partial}{\partial t}(E^2 + H^2) = -\frac{4\pi}{c}\mathbf{j}\cdot\mathbf{E} - \operatorname{div}(\mathbf{E}\times\mathbf{H})$$

or

$$\frac{\partial}{\partial t}\left(\frac{E^2 + H^2}{8\pi}\right) = -\mathbf{j}\cdot\mathbf{E} - \operatorname{div}\mathbf{S}. \tag{57.1}$$

The vector

$$\mathbf{S} = \frac{c}{4\pi}\mathbf{E}\times\mathbf{H} \tag{57.2}$$

is called the *Poynting vector*.

We integrate (57.1) over a volume and apply Gauss' theorem to the second term on the right. Then we obtain

$$\frac{\partial}{\partial t}\int \frac{E^2 + H^2}{8\pi}\,\mathrm{d}V = -\int \mathbf{j}\cdot\mathbf{E}\,\mathrm{d}V - \oint \mathbf{S}\cdot\mathrm{d}\mathbf{f}. \tag{57.3}$$

If the integral extends over all space, then the surface integral vanishes (the field is zero at infinity). Furthermore, we can express the integral $\int \mathbf{j}\cdot\mathbf{E}\,\mathrm{d}V$ as a sum $\Sigma e\mathbf{v}\cdot\mathbf{E}$ over all the charges, and substitute from (44.7)

$$e\mathbf{v}\cdot\mathbf{E} = \frac{\mathrm{d}}{\mathrm{d}t}\mathscr{E}_{\mathrm{kin}}.$$

Then (57.3) becomes

$$\frac{\mathrm{d}}{\mathrm{d}t}\left\{\int \frac{E^2+H^2}{8\pi}\,\mathrm{d}V+\sum \mathscr{E}_{\mathrm{kin}}\right\} = 0. \tag{57.4}$$

Thus for the closed system consisting of the electromagnetic field and particles present in it, the quantity in braces in this equation is conserved. The second term in this expression is the kinetic energy (including the rest energy of all the particles; see the footnote on p. 157); the first term is consequently the energy of the field itself. We can therefore call the quantity

$$W = \frac{E^2+H^2}{8\pi} \tag{57.5}$$

the *energy density* of the electromagnetic field; it is the energy per unit volume of the field.

If we integrate over a finite volume, the surface integral in (57.3) generally does not vanish, so that we can write the equation in the form

$$\frac{\partial}{\partial t}\left\{\int \frac{E^2+H^2}{8\pi}\,\mathrm{d}V+\sum \mathscr{E}_{\mathrm{kin}}\right\} = -\oint \mathbf{S}\cdot\mathrm{d}\mathbf{f}, \tag{57.6}$$

where now the second term in the braces is summed only over the particles present in the volume under consideration. On the left stands the change in the total energy of field and particles per unit time. Therefore the integral $\oint \mathbf{S}\cdot\mathrm{d}\mathbf{f}$ must be interpreted as the flux of field energy across the surface bounding the given volume, so that the Poynting vector $\mathbf{S}$ is this flux density—the amount of field energy passing through unit area of the surface in unit time.

§58. Momentum density and momentum flux

In addition to energy, the electromagnetic field also possesses momentum, distributed in space with a definite density. The expression for this density in terms of the field strengths can be found by a derivation similar to the one in the preceding section.

Let us calculate the time derivative of the integral

$$\int \frac{1}{4\pi c}\,\mathbf{E}\times\mathbf{H}\,\mathrm{d}V.$$

Carrying out the differentiation under the integral sign and replacing the derivatives $\partial\mathbf{E}/\partial t$ and $\partial\mathbf{H}/\partial t$ by using Maxwell's equations, we get

$$\frac{\partial}{\partial t}\int\frac{\mathbf{E}\times\mathbf{H}}{4\pi c}\,\mathrm{d}V = \frac{1}{4\pi c}\int\mathbf{E}\times\frac{\partial\mathbf{H}}{\partial t}\,\mathrm{d}V+\frac{1}{4\pi c}\int\frac{\partial\mathbf{E}}{\partial t}\times\mathbf{H}\,\mathrm{d}V =$$

$$= -\frac{1}{4\pi}\int\left\{\mathbf{E}\times\operatorname{curl}\mathbf{E}+\mathbf{H}\times\operatorname{curl}\mathbf{H}\right\}\mathrm{d}V-\frac{1}{c}\int\mathbf{j}\times\mathbf{H}\,\mathrm{d}V.$$

In the first integral we transform the integrand by using the formula from vector analysis

$$\nabla(\mathbf{a}\cdot\mathbf{b}) = \mathbf{a}\times\operatorname{curl}\mathbf{b}+\mathbf{b}\times\operatorname{curl}\mathbf{a}+(\mathbf{a}\cdot\nabla)\mathbf{b}+(\mathbf{b}\cdot\nabla)\mathbf{a},$$

according to which we have

$$\mathbf{E}\times\operatorname{curl}\mathbf{E} = \tfrac{1}{2}\nabla E^2-(\mathbf{E}\cdot\nabla)\mathbf{E}.$$

We also write

$$(\mathbf{E}\cdot\nabla)\mathbf{E} = (\nabla\cdot\mathbf{E})\mathbf{E}-\mathbf{E}(\nabla\cdot\mathbf{E}),$$

where, in the term $(\nabla\cdot\mathbf{E})\mathbf{E}$, it is assumed that the operator ∇ acts on both of the factors that follow it. Finally, noting that, according to Maxwell's equation (56.4), $\nabla\cdot\mathbf{E}\equiv\operatorname{div}\mathbf{E} = 4\pi\varrho$, we write

$$\mathbf{E}\times\operatorname{curl}\mathbf{E} = \tfrac{1}{2}\nabla E^2-(\nabla\cdot\mathbf{E})\mathbf{E}+4\pi\varrho\,\mathbf{E}.$$

The product $\mathbf{H}\times\operatorname{curl}\mathbf{H}$ is transformed similarly, but since $\operatorname{div}\mathbf{H} = 0$,

$$\mathbf{H}\times\operatorname{curl}\mathbf{H} = \tfrac{1}{2}\nabla H^2-(\nabla\cdot\mathbf{H})\mathbf{H}.$$

Thus,

$$\frac{\partial}{\partial t}\int \frac{\mathbf{E}\times\mathbf{H}}{4\pi c}\,\mathrm{d}V = -\int \frac{1}{4\pi}\left\{\frac{1}{2}\nabla(E^2+H^2)-(\nabla\cdot\mathbf{E})\mathbf{E}-(\nabla\cdot\mathbf{H})\mathbf{H}\right\}\mathrm{d}V-$$
$$-\int\left\{\varrho\mathbf{E}+\frac{1}{c}\,\mathbf{j}\times\mathbf{H}\right\}\mathrm{d}V. \qquad (58.1)$$

In the first integral the operator ∇ acts on all terms following it in the integrand. According to vector analysis (the general formulation of Gauss' theorem), this integral is transformed into a surface integral by writing the surface element $\mathrm{d}\mathbf{f}$ in place of the operator $\mathrm{d}V\cdot\nabla$. In the second integral, in which the charge density and current appear, we change to the discrete form in terms of a sum over the point charges located within the given volume. As a result, (58.1) is rewritten in the form

$$\frac{\partial}{\partial t}\int \frac{\mathbf{E}\times\mathbf{H}}{4\pi c}\,\mathrm{d}V = -\oint \frac{1}{4\pi}\left\{\frac{1}{2}(E^2+H^2)\,\mathrm{d}\mathbf{f}-\mathbf{E}(\mathbf{E}\cdot\mathrm{d}\mathbf{f})-\mathbf{H}(\mathbf{H}\cdot\mathrm{d}\mathbf{f})\right\}-$$
$$-\sum e\left(\mathbf{E}+\frac{1}{c}\,\mathbf{v}\times\mathbf{H}\right). \qquad (58.2)$$

If the integral is extended over all space, the integral over the (infinitely distant) surface vanishes. The expression under the summation sign in (58.2) is the force acting on the charge. According to the equation of motion (44.5) it can be replaced by the derivative $\mathrm{d}\mathbf{p}/\mathrm{d}t$ of the momentum of the particle. Then (58.2) can be written in the form

$$\frac{\partial}{\partial t}\left\{\int \frac{\mathbf{E}\times\mathbf{H}}{4\pi c}\,\mathrm{d}V+\sum \mathbf{p}\right\} = 0. \qquad (58.3)$$

It is obviously a statement of the law of conservation of the total momentum of the system of particles plus field. The first term in the braces is thus the momentum of the electromagnetic field, while the integrand in it can be regarded as the momentum density; we denote it by $\mathbf{P}^{(em)}$:

$$\mathbf{P}^{(em)} = \mathbf{E}\times\mathbf{H}/4\pi c = \mathbf{S}/c^2. \tag{58.4}$$

We point out that (except for a factor $1/c^2$) the momentum density coincides with the flux density of the field energy.

But if the integration on the left-hand side of (58.2) is extended over some finite volume of field, the surface integral is not equal to zero. We write it in a more compact form, introducing the three-dimensional tensor

$$\sigma_{ik} = \frac{1}{4\pi}\left\{\frac{1}{2}(E^2+H^2)\delta_{ik} - E_iE_k - H_iH_k\right\}. \tag{58.5}$$

In expanded form its components are

$$\sigma_{xx} = \frac{1}{8\pi}(E_y^2+E_z^2-E_x^2+H_y^2+H_z^2-H_x^2),$$

$$\sigma_{xy} = -\frac{1}{4\pi}(E_xE_y+H_xH_y)$$

etc.

The integrand in the surface integral in (58.2) is a vector; using the tensor (58.5), we can write its ith component as $\sigma_{ik}\,\mathrm{d}f_k$. Thus the vector equation of momentum conservation (58.2), when written in this compact form, is

$$\frac{\partial}{\partial t}\left\{\int P_i^{(em)}\,\mathrm{d}V + \sum p_i\right\} = -\oint \sigma_{ik}\,\mathrm{d}f_k. \tag{58.6}$$

We then see that the integral on the right-hand side represents the flux of field momentum out of the volume. The product $\sigma_{ik}\,\mathrm{d}f_k$ is the flux of momentum through the surface element $\mathbf{df}$. By definition, the vector $\mathbf{df}$ is directed along the external normal to the surface. If we denote the unit normal vector by $\mathbf{N}$, then $\mathbf{df} = \mathbf{N}\,\mathrm{d}f$, and

$$\sigma_{ik}\,\mathrm{d}f_k = \sigma_{ik}N_k\,\mathrm{d}f;$$

we see that the vector with components $\sigma_{ik}N_k$ is the momentum flux density in the direction **N**, i.e. the flux through unit area perpendicular to **N**. Substituting for σ_{ik} from (58.5), we find that this vector is

$$\frac{1}{4\pi}\left\{\frac{1}{2}(E^2+H^2)\mathbf{N}-\mathbf{E}(\mathbf{N}\cdot\mathbf{E})-\mathbf{H}(\mathbf{N}\cdot\mathbf{H})\right\}. \tag{58.7}$$

The tensor σ_{ik} is called the ***Maxwell stress tensor***. According to our discussion, the component σ_{ik} is the flux density of the ith component of the momentum in the direction of the x^k axis. We note that the stress tensor, as we see from (58.5), is symmetric ($\sigma_{ik} = \sigma_{ki}$).

CHAPTER 12

CONSTANT ELECTROMAGNETIC FIELDS

§59. Coulomb's law

For a constant electric (*electrostatic*) field, Maxwell's equations have the form

$$\operatorname{div} \mathbf{E} = 4\pi\varrho, \tag{59.1}$$

$$\operatorname{curl} \mathbf{E} = 0. \tag{59.2}$$

The electric field **E** is expressed in terms of the scalar potential alone by the relation

$$\mathbf{E} = -\operatorname{grad} \phi. \tag{59.3}$$

Substituting (59.3) in (59.1), we get the equation which is satisfied by the potential of a constant electric field:

$$\triangle\phi = -4\pi\varrho. \tag{59.4}$$

This equation is called *Poisson's equation*. In vacuum, i.e., for $\varrho = 0$, the potential satisfies *Laplace's equation*

$$\triangle\phi = 0. \tag{59.5}$$

From the last equation it follows, in particular, that the potential of the electric field can nowhere have a maximum or a minimum. For in order that ϕ have an extreme value, it would be necessary that the first derivatives of ϕ with respect to the coordinates be zero, and that the second derivatives $\partial^2\phi/\partial x^2$, $\partial^2\phi/\partial y^2$, $\partial^2\phi/\partial z^2$ all have the same sign. The last is impossible, since in that case (59.5) could not be satisfied.

We now determine the field produced by a point charge. From symmetry considerations, it is clear that it is directed along the radius-vector from the point at which the charge e is located. From the same considerations it is clear that the value E of the field depends only on the distance R from the charge. To find this absolute value, we apply equation (59.1) in the integral form (56.5). The flux of the electric field through a spherical surface of radius R circumscribed around the charge e is equal to $4\pi R^2 E$; this flux must equal $4\pi e$. From this we get

$$E = e/R^2.$$

In vector notation,

$$\mathbf{E} = e\mathbf{R}/R^3. \tag{59.6}$$

Thus the field produced by a point charge is inversely proportional to the square of the distance from the charge. This is *Coulomb's law*. The potential of this field is, clearly,

$$\phi = e/R. \tag{59.7}$$

If we have a system of charges, then the field produced by this system is equal, according to the principle of superposition, to the sum of the fields produced by each of the charges individually. In particular, the potential of such a field is

$$\phi = \sum_a \frac{e_a}{R_a}, \tag{59.8}$$

where R_a is the distance from the charge e_a to the point at which we are determining the potential. If we introduce the charge density ϱ, this formula takes on the form

$$\phi = \int \frac{\varrho}{R}\, \mathrm{d}V, \tag{59.9}$$

where R is the distance from the volume element $\mathrm{d}V$ to the given field point.

We note a mathematical relation which is obtained from (59.4) by substituting the values of ϱ and ϕ for a point charge, i.e. $\varrho = e\delta(\mathbf{R})$ and $\phi = e/R$. We then find

$$\triangle(1/R) = -4\pi\delta(\mathbf{R}). \tag{59.10}$$

§60. Electrostatic energy of charges

Let us determine the potential energy of a system of charges. We start from the energy of the field, that is, from the expression (57.5) for the energy density. The energy of the system of charges must be equal to

$$U = \frac{1}{8\pi}\int E^2 \, \mathrm{d}V,$$

where $\mathbf{E}$ is the field produced by these charges, and the integral goes over all space. Substituting $\mathbf{E} = -\operatorname{grad}\phi$, U can be changed to the following form:

$$U = -\frac{1}{8\pi}\int \mathbf{E}\cdot\operatorname{grad}\phi \, \mathrm{d}V = -\frac{1}{8\pi}\int \operatorname{div}(\mathbf{E}\phi)\,\mathrm{d}V + \frac{1}{8\pi}\int \phi \operatorname{div}\mathbf{E}\,\mathrm{d}V.$$

According to Gauss' theorem, the first integral is equal to the integral of $\mathbf{E}\phi$ over the surface bounding the volume of integration, but since the integral is taken over all space and since the field is zero at infinity, this integral vanishes. Substituting in the second integral $\operatorname{div}\mathbf{E} = 4\pi\varrho$, we find the following expression for the energy of a system of charges:

$$U = \tfrac{1}{2}\int \varrho\phi \, \mathrm{d}V. \tag{60.1}$$

For a system of point charges, e_a, we can write in place of the integral a sum over the charges

$$U = \tfrac{1}{2}\sum_a e_a\phi_a, \tag{60.2}$$

where ϕ_a is the potential of the field produced by all the charges, at the point where the charge e_a is located.

According to formula (59.8), the potentials ϕ_a are equal to

$$\phi_a = \sum_b \frac{e_b}{R_{ab}},$$

where R_{ab} is the distance between the charges e_a, e_b. For a system of point charges, this expression contains an infinite term coming from the potential of the self-field of the charge e_a (the term of the sum with $b = a$, in which $R_{aa} = 0$). Accordingly, there is an infinite constant in the energy (60.2), which is independent of the separation of the charges. This part of the energy—the *self-energy* of the charges—is physically meaningless (see below) and should be dropped. There then remains only the interaction energy of the charges that depends on their separations. It is equal to

$$U' = \tfrac{1}{2} \sum_a e_a \phi'_a, \tag{60.3}$$

where

$$\phi'_a = \sum_{b \neq a} \frac{e_b}{R_{ab}} \tag{60.4}$$

is the potential at the point of location of e_a, produced by all the charges other than e_a. In other words, we can write

$$U' = \frac{1}{2} \sum_{a \neq b} \frac{e_a e_b}{R_{ab}}. \tag{60.5}$$

In particular, the energy of interaction of two charges is

$$U' = \frac{e_1 e_2}{R_{12}}. \tag{60.6}$$

Let us look at the infinite self-energy of an elementary charged particle that was mentioned above. It arose as a result of considering the particles to be points. But such a treatment is unavoidable in classical (non-quantum) relativistic theory because of the fundamental principles of the theory of relativity. For, when in classical theory we speak of an elementary particle, we mean a particle whose mechanical state is completely described by giving its coordinates and its velocity of motion as a whole. If such a particle were extended, then it would, in any case, have to be regarded as an absolutely rigid body (i.e. a body incapable of deformation), since the very concept of deformation is related to the possibility of independent displacements of individual parts of the body. But in relativistic mechanics the existence of absolutely rigid bodies is impossible, as can be seen from the following arguments.

Suppose that a rigid body is set in motion by an external force acting at one of its points. If the body were absolutely rigid, then all of its points would have to start moving simultaneously with the point that was subjected to the force; otherwise the body would deform. But, because there is a limiting velocity of propagation of interactions, the force is transmitted from the initial point of application to other points with a finite velocity, and therefore all the points of the body cannot begin to move simultaneously.

Thus, according to electrodynamics the electron should have an infinite self-energy, and consequently an infinite mass. The physical meaninglessness of this result shows that electrodynamics as a logically closed physical theory becomes self-contradictory when we go to sufficiently small distances. We may ask about the order of magnitude of these distances. We can answer this question by noting that we should obtain a value for the electromagnetic self-energy of the electron that is of the order of the rest energy mc^2. If now we consider the electron as having a certain size r_e, its self-potential energy would be of order e^2/r_e. From the requirement that these quantities be of the same order of magnitude, $e^2/r_e \sim mc^2$, we find

$$r_e \sim e^2/mc^2. \tag{60.7}$$

This dimension (called the "radius" of the electron) determines the limit of applicability of electrodynamics to the electron which already follows from the fundamental principles of electrodynamics. We must, however, keep in mind that actually the limit of applicability of the classical electrodynamics which is presented here lies much higher, because of quantum phenomena.†

---

† Quantum effects begin to be important at distances of order $\hbar/mc$, where $\hbar$ is Planck's constant.

§61. The field of a uniformly moving charge

We determine the field produced by a charge e, moving uniformly with velocity V. We call the laboratory system the frame K; the frame of reference moving with the charge is the frame K'. Let the charge be located at the origin of coordinates of the frame K'. The frame K' moves relative to K along the axis X; the axes Y and Z are parallel to Y' and Z'. At the time $t = 0$ the origins of the two systems coincide. The coordinates of the charge in the frame K are consequently $x = Vt$, $y = z = 0$.

In the K' system the electric field is constant:

$$\mathbf{E}' = \frac{e\mathbf{R}'}{R'^3}, \tag{61.1}$$

and the magnetic field is absent. Transforming to the frame K, using (50.5), we have

$$E_x = \frac{ex'}{R'^3}, \quad E_y = \frac{ey'}{R'^3\sqrt{[1-(V^2/c^2)]}}, \quad E_z = \frac{ez'}{R'^3\sqrt{[1-(V^2/c^2)]}}. \tag{61.2}$$

We must now express R', x', y', z' in terms of the coordinates x, y, z in the frame K. According to the formulae for the Lorentz transformation,

$$x' = \frac{x-Vt}{\sqrt{[1-(V^2/c^2)]}}, \quad y' = y, \quad z' = z,$$

so that

$$R'^2 = \frac{R^{*2}}{1-(V^2/c^2)}, \tag{61.3}$$

where

$$R^{*2} = (x-Vt)^2+[1-(V^2/c^2)](y^2+z^2). \tag{61.4}$$

Substituting these expressions in (61.2), we obtain

$$\mathbf{E} = \left(1-\frac{V^2}{c^2}\right)\frac{e\mathbf{R}}{R^{*3}}, \tag{61.5}$$

where $\mathbf{R}$ is the radius vector from the charge e to the field point with coordinates x, y, z (its components are $x-Vt$, y, z).

This expression for **E** can be written in another form by introducing the angle θ between the direction of motion and the radius vector **R**. It is clear that $y^2+z^2 = R^2 \sin^2 \theta$, and therefore

$$R^{*2} = R^2[1-(V^2/c^2)\sin^2\theta].$$

Then we have for **E**

$$\mathbf{E} = \frac{e\mathbf{R}}{R^3}\frac{1-V^2/c^2}{[1-(V^2/c^2)\sin^2\theta]^{3/2}}. \tag{61.6}$$

For a fixed distance R from the charge, the value of the field E increases as θ increases from 0 to $\pi/2$ (or as θ decreases from π to $\pi/2$). The field along the direction of motion ($\theta = 0, \pi$) has the smallest value; it is equal to

$$E_{\|} = (e/R^2)[1-(V^2/c^2)].$$

The largest field is that perpendicular to the velocity ($\theta = \pi/2$), equal to

$$E_{\perp} = \frac{e}{R^2}\frac{1}{\sqrt{[1-(V^2/c^2)]}}.$$

We note that, as the velocity increases, the field $E_{\|}$ decreases, while $E_{\perp}$ increases. We can describe this pictorially by saying that the electric field of a moving charge is "contracted" in the direction of motion. For velocities V close to the velocity of light, the denominator in formula (61.6) is close to zero in a narrow interval of values θ around the value $\theta = \pi/2$. The "width" of this interval is, in order of magnitude,

$$\Delta\theta \sim \sqrt{[1-(V^2/c^2)]}.$$

Thus the electric field of a rapidly moving charge at a given distance from it is significantly different from zero only in a narrow range of angles in the neighbourhood of the equatorial plane, and the width of this interval decreases with increasing V as $\sqrt{[1-(V^2/c^2)]}$.

The magnetic field in the frame K is

$$\mathbf{H} = \frac{1}{c}\mathbf{V}\times\mathbf{E} \tag{61.7}$$

[see (50.7)]. In particular, for $V \ll c$ the electric field is given approximately by the usual formula for Coulomb's law, $\mathbf{E} = e\mathbf{R}/R^3$, and the magnetic field is

$$\mathbf{H} = \frac{e}{c}\frac{\mathbf{V}\times\mathbf{R}}{R^3}. \tag{61.8}$$

PROBLEM

Determine the force (in the frame K) between two charges moving with the same velocity $\mathbf{V}$.

SOLUTION. We shall determine the force $\mathbf{F}$ by computing the force acting on one of the charges (e_1) in the field produced by the other (e_2). Using (61.7), we have

$$\mathbf{F} = e_1\mathbf{E}_2+\frac{e_1}{c}\mathbf{V}\times\mathbf{H}_2 = e_1\left(1-\frac{V^2}{c^2}\right)\mathbf{E}_2+\frac{e_1}{c^2}\mathbf{V}(\mathbf{V}\cdot\mathbf{E}_2).$$

Substituting for $\mathbf{E}_2$ from (61.6), we get for the components of the force in the direction of motion (F_x) and perpendicular to it (F_y)

$$F_x = \frac{e_1e_2}{R^2}\frac{[1-(V^2/c^2)]\cos\theta}{[1-(V^2/c^2)\sin^2\theta]^{3/2}}, \quad F_y = \frac{e_1e_2}{R^2}\frac{[1-(V^2/c^2)]^2\sin\theta}{[1-(V^2/c^2)\sin^2\theta]^{3/2}},$$

where $\mathbf{R}$ is the radius vector from e_2 to e_1, and θ is the angle between $\mathbf{R}$ and $\mathbf{V}$.

§62. The dipole moment

We consider the field produced by a system of charges at large distances, that is, at distances large compared with the dimensions of the system.

We introduce a coordinate system with origin anywhere within the system of charges. Let the position vectors of the various charges be $\mathbf{r}_a$. The potential of the field produced by all the charges at the point having the position vector $\mathbf{R}_0$ is

$$\phi = \sum_a \frac{e_a}{|\mathbf{R}_0-\mathbf{r}_a|} \tag{62.1}$$

(the summation goes over all charges); here $\mathbf{R}_0-\mathbf{r}_a$ are the radius vectors from the charges e_a to the point where we are finding the potential.

We must investigate this expression for large $\mathbf{R}_0$ ($\mathbf{R}_0 \gg \mathbf{r}_a$). To do this, we expand it in powers of $\mathbf{r}_a/R_0$, using the formula

$$f(\mathbf{R}_0-\mathbf{r}) \approx f(\mathbf{R}_0)-\mathbf{r}\cdot\operatorname{grad} f(\mathbf{R}_0)$$

(in the grad, the differentiation is with respect to the coordinates of the vector $\mathbf{R}_0$). To terms of first order,

$$\phi = \frac{\sum e_a}{R_0} - \sum e_a\mathbf{r}_a\cdot\operatorname{grad}\frac{1}{R_0}. \tag{62.2}$$

The sum

$$\mathbf{d} = \sum e_a\mathbf{r}_a \tag{62.3}$$

is called the *dipole moment* of the system of charges.

It is important to note that, if the sum of all the charges, Σe_a, is zero, then the dipole moment does not depend on the choice of the origin of coordinates, for the position vectors $\mathbf{r}_a$ and $\mathbf{r}'_a$ of one and the same charge in two different coordinate systems are related by

$$\mathbf{r}'_a = \mathbf{r}_a+\mathbf{a},$$

where $\mathbf{a}$ is some constant vector. Therefore if $\Sigma e_a = 0$, the dipole moment is the same in both systems:

$$\mathbf{d}' = \sum e_a\mathbf{r}'_a = \sum e_a\mathbf{r}_a+\mathbf{a}\sum e_a = \mathbf{d}.$$

In particular, for a system of two charges with opposite signs ($\pm e$) the dipole moment $\mathbf{d} = e\mathbf{r}$, where $\mathbf{r}$ is the radius vector from the charge $-e$ to the charge $+e$.

If the total charge of the system is zero, then the potential of the field of this system at large distances is

$$\phi = -\mathbf{d}\cdot\nabla\frac{1}{R_0} = \frac{\mathbf{d}\cdot\mathbf{R}_0}{R_0^3}. \tag{62.4}$$

The field intensity is

$$\mathbf{E} = -\operatorname{grad}\frac{\mathbf{d}\cdot\mathbf{R}_0}{R_0^3} = -\frac{1}{R_0^3}\operatorname{grad}(\mathbf{d}\cdot\mathbf{R}_0)-(\mathbf{d}\cdot\mathbf{R}_0)\operatorname{grad}\frac{1}{R_0^3},$$

or finally,

$$\mathbf{E} = \frac{3(\mathbf{n}\cdot\mathbf{d})\mathbf{n}-\mathbf{d}}{R_0^3}, \tag{62.5}$$

where **n** is a unit vector along $\mathbf{R}_0$.

Thus the potential of the field at large distances produced by a system of charges with total charge equal to zero is inversely proportional to the square of the distance, and the field intensity is inversely proportional to the cube of the distance. This field has axial symmetry around the direction of **d**. In a plane passing through this direction (which we choose as the z axis), the components of the vector **E** are

$$E_z = d\frac{3\cos^2\theta - 1}{R_0^3}, \quad E_x = d\frac{3\sin\theta\cos\theta}{R_0^3}. \tag{62.6}$$

The radial and tangential components in this plane are

$$E_R = d\frac{2\cos\theta}{R_0^3}, \quad E_\theta = -d\frac{\sin\theta}{R_0^3}. \tag{62.7}$$

§63. The quadrupole moment

In the expansion of the potential in powers of $1/R_0$,

$$\phi = \phi^{(0)} + \phi^{(1)} + \phi^{(2)} + \ldots, \tag{63.1}$$

the term $\phi^{(n)}$ is proportional to $1/R_0^{n+1}$. We saw that the first term, $\phi^{(0)}$, is determined by the sum of all the charges; the second term, $\phi^{(1)}$, called the dipole potential of the system, is determined by the dipole moment of the system.

The third term in the expansion is

$$\phi^{(2)} = \tfrac{1}{2}\sum e x_i x_k \frac{\partial^2}{\partial X_i \partial X_k}\left(\frac{1}{R_0}\right), \tag{63.2}$$

where the sum goes over all charges; we here omit the suffix numbering the charges; x_i are the components of the vector **r**, and X_i those of the vector $\mathbf{R}_0$. This part of the potential is usually called the *quadrupole* *potential.* If the sum of the charges and the dipole moment of the system are both equal to zero, the expansion begins with $\phi^{(2)}$.

In the expression (63.2) there enter the six quantities $\Sigma e x_i x_k$. However, it is easy to see that the field depends not on six independent quantities, but only on five. This follows from the fact that the function $1/R_0$ satisfies Laplace's equation, that is,

$$\triangle\left(\frac{1}{R_0}\right) \equiv \delta_{ik}\frac{\partial^2}{\partial X_i\,\partial X_k}\left(\frac{1}{R_0}\right) = 0.$$

We can therefore write $\phi^{(2)}$ in the form

$$\phi^{(2)} = \tfrac{1}{2}\sum e\left(x_i x_k - \tfrac{1}{3}r^2\delta_{ik}\right)\frac{\partial^2}{\partial X_i\,\partial X_k}\left(\frac{1}{R_0}\right).$$

The tensor

$$D_{ik} = \sum e(3x_i x_k - r^2\delta_{ik}) \tag{63.3}$$

is called the *quadrupole moment* tensor of the system. From the definition of D_{ik} it is clear that the sum of its diagonal elements is zero:

$$D_{ii} = 0. \tag{63.4}$$

Therefore the symmetric tensor D_{ik} has only five independent components. With the aid of D_{ik} we can write

$$\phi^{(2)} = \frac{D_{ik}}{6}\,\frac{\partial^2}{\partial X_i\,\partial X_k}\left(\frac{1}{R_0}\right), \tag{63.5}$$

or, performing the differentiation,

$$\frac{\partial^2}{\partial X_i\,\partial X_k}\left(\frac{1}{R_0}\right) = \frac{3X_i X_k}{R_0^5} - \frac{\delta_{ik}}{R_0^3},$$

and using the fact that $\delta_{ik}D_{ik} = D_{ii} = 0$,

$$\phi^{(2)} = \frac{D_{ik}n_i n_k}{2R_0^3}. \tag{63.6}$$

Like every symmetric three-dimensional tensor, the tensor D_{ik} can be brought to principal axes. Because of (63.4), in general only two of the three principal values will be independent. If it happens that the system of charges is symmetric around some axis (the z axis)† then this axis must be one of the principal axes of the tensor D_{ik}; the location of the other two axes in the x, y plane is arbitrary, and the three principal values are related to one another:

$$D_{xx} = D_{yy} = -\tfrac{1}{2}D_{zz}. \tag{63.7}$$

Denoting the component D_{zz} by D (in this case it is simply called the quadrupole moment), we get for the potential

$$\phi^{(2)} = \frac{D}{4R_0^3}(3\cos^2\theta - 1), \tag{63.8}$$

where θ is the angle between $\mathbf{R}_0$ and the z axis.

Just as we did for the dipole moment in the preceding section, we can easily show that the quadrupole moment of a system does not depend on the choice of the coordinate origin, if both the total charge and the dipole moment of the system are equal to zero.

In similar fashion we could also write the succeeding terms of the expansion (63.1). The lth term of the expansion defines a tensor (which is called the 2^l-pole moment tensor) of rank l, symmetric in all its indices and vanishing when contracted on any pair of indices; it can be shown that such a tensor has $2l+1$ independent components.

PROBLEM

Determine the quadrupole moment of a uniformly charged ellipsoid with respect to its centre.

SOLUTION. Replacing the summation in (63.3) by an integration over the volume of the ellipsoid, we have

$$D_{xx} = \varrho \iiint (2x^2 - y^2 - z^2)\, dx\, dy\, dz, \text{ etc.}$$

The integration over the volume of the ellipsoid can be reduced to integration over the volume of the unit sphere as was done in §25, Problem 2(e). As a result we obtain

$$D_{xx} = \tfrac{1}{5}e(2a^2 - b^2 - c^2), \quad D_{yy} = \tfrac{1}{5}e(2b^2 - a^2 - c^2),$$
$$D_{zz} = \tfrac{1}{5}e(2c^2 - a^2 - b^2),$$

where $e = (4\pi/3)abc\varrho$ is the total charge of the ellipsoid.

---

† We are assuming a symmetry axis of any order higher than the second.

§64. System of charges in an external field

We now consider a system of charges located in an external electric field. We designate the potential of this external field by $\phi(\mathbf{r})$. The potential energy of each of the charges is $e_a\phi(\mathbf{r}_a)$, and the total potential energy of the system is

$$U = \sum_a e_a\phi(\mathbf{r}_a). \tag{64.1}$$

We again introduce a coordinate system with its origin anywhere within the system of charges; $\mathbf{r}_a$ is the position of the charge e_a in these coordinates.

Let us assume that the external field changes slowly over the region of the system of charges, i.e. is *quasi-uniform* with respect to the system. Then we can expand the energy U in powers of $\mathbf{r}_a$. In this expansion,

$$U = U^{(0)} + U^{(1)} + U^{(2)} + \ldots, \tag{64.2}$$

the first term is

$$U^{(0)} = \phi_0 \sum e_a, \tag{64.3}$$

where ϕ_0 is the value of the potential at the origin. In this approximation, the energy of the system is the same as it would be if all the charges were located at one point.

The second term in the expansion is

$$U^{(1)} = (\text{grad}\,\phi)_0 \cdot \sum e_a \mathbf{r}_a.$$

Introducing the field intensity $\mathbf{E}_0$ at the origin and the dipole moment $\mathbf{d}$ of the system, we have

$$U^{(1)} = -\mathbf{d}\cdot\mathbf{E}_0. \tag{64.4}$$

The total force acting on the system in the external quasi-uniform field is, to the order we are considering,

$$\mathbf{F} = \mathbf{E}_0 \sum e_a + (\nabla \mathbf{d}\cdot\mathbf{E})_0.$$

If the total charge is zero, the first term vanishes, and

$$\mathbf{F} = (\mathbf{d}\cdot\nabla)\mathbf{E}, \tag{64.5}$$

i.e. the force is determined by the derivatives of the field intensity (taken at the origin). The total moment of the forces acting on the system is

$$\mathbf{K} = \sum (\mathbf{r}_a \times e_a \mathbf{E}_0) = \mathbf{d} \times \mathbf{E}_0, \tag{64.6}$$

i.e. it is determined by the field intensity itself.

Let us assume that there are two systems, each having total charge zero, and with dipole moments $\mathbf{d}_1$ and $\mathbf{d}_2$, respectively. Their mutual distance R is assumed to be large in comparison with their internal dimensions. Let us determine their potential energy of interaction, U. To do this we regard one of the systems as being in the field of the other. Then

$$U = -\mathbf{d}_2 \cdot \mathbf{E}_1,$$

where $\mathbf{E}_1$ is the field of the first system. Substituting (62.5) for $\mathbf{E}_1$, we find

$$U = \frac{(\mathbf{d}_1 \cdot \mathbf{d}_2) - 3(\mathbf{d}_1 \cdot \mathbf{n})(\mathbf{d}_2 \cdot \mathbf{n})}{R^3}, \tag{64.7}$$

where $\mathbf{n}$ is a unit vector from one system to the other.

For the case where one of the systems has a total charge different from zero (and equal to e), we obtain similarly

$$U = e\frac{\mathbf{d} \cdot \mathbf{n}}{R^2}, \tag{64.8}$$

where $\mathbf{n}$ is the unit vector directed from the dipole to the charge.

The next term in the expansion (64.2) is

$$U^{(2)} = \tfrac{1}{2}\sum e x_i x_k \frac{\partial^2 \phi_0}{\partial x_i \partial x_k}.$$

Here, as in §63, we omit the suffix numbering the charge; the values of the second derivatives of the potential are taken at the origin; but the potential ϕ satisfies Laplace's equation,

$$\frac{\partial^2 \phi}{\partial x^2} = \delta_{ik}\frac{\partial^2 \phi}{\partial x_i \partial x_k} = 0.$$

Therefore we can write

$$U^{(2)} = \tfrac{1}{2} \frac{\partial^2 \phi_0}{\partial x_i \, \partial x_k} \sum e(x_i x_k - \tfrac{1}{3}\, \delta_{ik} r^2)$$

or, finally,

$$U^{(2)} = \frac{D_{ik}}{6} \frac{\partial^2 \phi_0}{\partial x_i \, \partial x_k} . \tag{64.9}$$

§65. Constant magnetic field

Let us consider the magnetic field produced by charges which execute a finite motion, in which the particles are always within a finite region of space and the momenta also always remain finite. Such a motion has a "stationary" character, and it is of interest to consider the time average magnetic field $\overline{\mathbf{H}}$, produced by the charges; this field will now be a function only of the coordinates and not of the time, that is, it will be constant.

In order to find equations for the average magnetic field $\overline{\mathbf{H}}$, we take the time average of Maxwell's equations

$$\operatorname{div} \mathbf{H} = 0, \quad \operatorname{curl} \mathbf{H} = \frac{1}{c} \frac{\partial \mathbf{E}}{\partial t} + \frac{4\pi}{c} \mathbf{j}.$$

The first of these gives simply

$$\operatorname{div} \overline{\mathbf{H}} = 0. \tag{65.1}$$

In the second equation the average value of the derivative $\partial \mathbf{E}/\partial t$, like the derivative of any quantity which varies over a finite range, is zero†.

† Suppose that f is such a quantity. Then the average value of the derivative $\mathrm{d}f/\mathrm{d}t$ over a time interval T is

$$\overline{\frac{\mathrm{d}f}{\mathrm{d}t}} = \frac{1}{T} \int_0^T \frac{\mathrm{d}f}{\mathrm{d}t} \mathrm{d}t = \frac{f(T) - f(0)}{T} .$$

Since $f(t)$ varies only over a finite range, as T increases without limit the average value tends to zero.

Therefore the second Maxwell's equation becomes

$$\operatorname{curl} \bar{\mathbf{H}} = \frac{4\pi}{c} \bar{\mathbf{j}}. \tag{65.2}$$

These two equations determine the constant field $\bar{\mathbf{H}}$.

We introduce the average vector potential $\bar{\mathbf{A}}$ in accordance with

$$\operatorname{curl} \bar{\mathbf{A}} = \bar{\mathbf{H}}.$$

We substitute this in equation (65.2), and find

$$\operatorname{grad} \operatorname{div} \bar{\mathbf{A}} - \triangle \bar{\mathbf{A}} = \frac{4\pi}{c} \bar{\mathbf{j}}.$$

But we know that the vector potential of a field is not uniquely defined, and we can impose an arbitrary auxiliary condition on it. On this basis, we choose the potential $\bar{\mathbf{A}}$ so that

$$\operatorname{div} \bar{\mathbf{A}} = 0. \tag{65.3}$$

Then the equation defining the vector potential of the constant magnetic field becomes

$$\triangle \bar{\mathbf{A}} = -\frac{4\pi}{c} \bar{\mathbf{j}}. \tag{65.4}$$

It is easy to find the solution of this equation by noting that (65.4) is completely analogous to Poisson's equation (59.4) for the scalar potential of a constant electric field, where in place of the charge density ϱ we here have the current density $\bar{\mathbf{j}}/c$. By analogy with the solution (59.9) of Poisson's equation, we can write

$$\bar{\mathbf{A}} = \frac{1}{c} \int \frac{\bar{\mathbf{j}}}{R} \, dV, \tag{65.5}$$

where R is the distance from the field point to the volume element dV.

In formula (65.5) we can go over from the integral to a sum over the charges, by substituting in place of $\mathbf{j}$ the product $\varrho \mathbf{v}$, and recalling that all the charges are point-like. In this we must keep in mind that in the integral (65.5), R is simply an integration variable, and is therefore not subject to the averaging process. If we write in place of the integral

$$\int \frac{\mathbf{j}}{R}\,\mathrm{d}V \quad \text{the sum} \quad \sum \frac{e_a \mathbf{v}_a}{R_a},$$

then R_a here are the positions of the various particles, which change during the motion of the charges. Therefore we must write

$$\bar{\mathbf{A}} = \frac{1}{c}\sum \overline{\frac{e_a \mathbf{v}_a}{R_a}}, \tag{65.6}$$

where we average the whole expression under the bar.

Knowing $\bar{\mathbf{A}}$, we can also find the magnetic field,

$$\bar{\mathbf{H}} = \operatorname{curl}\bar{\mathbf{A}} = \operatorname{curl}\frac{1}{c}\int \frac{\bar{\mathbf{j}}}{R}\,\mathrm{d}V.$$

The curl operator refers to the coordinates of the field point. Therefore the curl can be brought under the integral sign and $\bar{\mathbf{j}}$ can be treated as constant in the differentiation. Applying the well-known formula

$$\operatorname{curl} f\mathbf{a} = f \operatorname{curl}\mathbf{a} + \operatorname{grad} f \times \mathbf{a},$$

where f and $\mathbf{a}$ are an arbitrary scalar and vector, to the product $\bar{\mathbf{j}}\cdot 1/R$, we get

$$\operatorname{curl}\frac{\bar{\mathbf{j}}}{R} = \operatorname{grad}\frac{1}{R}\times\bar{\mathbf{j}} = \frac{\bar{\mathbf{j}}\times\mathbf{R}}{R^3},$$

and consequently,

$$\bar{\mathbf{H}} = \frac{1}{c}\int \frac{\bar{\mathbf{j}}\times\mathbf{R}}{R^3}\,\mathrm{d}V \tag{65.7}$$

(the radius vector $\mathbf{R}$ is directed from $\mathrm{d}V$ to the field point). This is *Biot and Savart's law.*

§66. Magnetic moments

Let us consider the average magnetic field produced by a system of charges in steady motion, at large distances from the system.

We introduce a coordinate system with its origin anywhere within the system of charges, just as we did in §62. Again we denote the position vectors of the various charges by $\mathbf{r}_a$, and the position vector of the point at which we calculate the field by $\mathbf{R}_0$. Then $\mathbf{R}_0-\mathbf{r}_a$ is the radius vector from the charge e_a to the field point. According to (65.6), we have for the vector potential

$$\bar{\mathbf{A}} = \frac{1}{c}\sum \overline{\frac{e_a\mathbf{v}_a}{|\mathbf{R}_0-\mathbf{r}_a|}}. \tag{66.1}$$

As in §62, we expand this expression in powers of $\mathbf{r}_a$. To terms of first order we have (omitting the suffix a)

$$\bar{\mathbf{A}} = \frac{1}{cR_0}\sum e\bar{\mathbf{v}} - \frac{1}{c}\sum \overline{e\mathbf{v}\left(\mathbf{r}\cdot\nabla\frac{1}{R_0}\right)}.$$

In the first term we can write

$$\sum e\bar{\mathbf{v}} = \overline{\frac{\mathrm{d}}{\mathrm{d}t}\sum e\mathbf{r}}.$$

But the average value of the derivative of a quantity varying within a finite interval (like $\Sigma e\mathbf{r}$) is zero. Thus there remains for $\bar{\mathbf{A}}$ the expression

$$\bar{\mathbf{A}} = -\frac{1}{c}\sum \overline{e\mathbf{v}\left(\mathbf{r}\cdot\nabla\frac{1}{R_0}\right)} = \frac{1}{cR_0^3}\sum \overline{e\mathbf{v}(\mathbf{r}\cdot\mathbf{R}_0)}.$$

We transform this expression as follows. Noting that $\mathbf{v} = \dot{\mathbf{r}}$, we can write (remembering that $\mathbf{R}_0$ is a constant vector)

$$\sum e(\mathbf{R}_0\cdot\mathbf{r})\mathbf{v} = \tfrac{1}{2}\frac{\mathrm{d}}{\mathrm{d}t}\sum e\mathbf{r}(\mathbf{r}\cdot\mathbf{R}_0) + \tfrac{1}{2}\sum e[\mathbf{v}(\mathbf{r}\cdot\mathbf{R}_0) - \mathbf{r}(\mathbf{v}\cdot\mathbf{R}_0)].$$

Upon substitution of this expression in $\bar{\mathbf{A}}$, the average of the first term (containing the time derivative) again goes to zero, and we get

$$\bar{\mathbf{A}} = \frac{1}{2cR_0^3}\sum \overline{e[\mathbf{v}(\mathbf{r}\cdot\mathbf{R}_0)-\mathbf{r}(\mathbf{v}\cdot\mathbf{R}_0)]}\,.$$

We introduce the vector

$$\mathbf{m} = \frac{1}{2c}\sum e\mathbf{r}\times\mathbf{v}, \tag{66.2}$$

which is called the *magnetic moment* of the system. Then we get for $\bar{\mathbf{A}}$

$$\bar{\mathbf{A}} = \frac{\overline{\mathbf{m}}\times\mathbf{R}_0}{R_0^3} = \nabla\frac{1}{R_0}\times\overline{\mathbf{m}}. \tag{66.3}$$

Knowing the vector potential, it is easy to find the magnetic field. With the aid of the formula

$$\operatorname{curl}(\mathbf{a}\times\mathbf{b}) = (\mathbf{b}\cdot\nabla)\mathbf{a}-(\mathbf{a}\cdot\nabla)\mathbf{b}+\mathbf{a}\operatorname{div}\mathbf{b}-\mathbf{b}\operatorname{div}\mathbf{a},$$

we find

$$\bar{\mathbf{H}} = \operatorname{curl}\left(\frac{\overline{\mathbf{m}}\times\mathbf{R}_0}{R_0^3}\right) = \overline{\mathbf{m}}\operatorname{div}\frac{\mathbf{R}_0}{R_0^3}-(\overline{\mathbf{m}}\cdot\nabla)\frac{\mathbf{R}_0}{R_0^3}\,.$$

Furthermore,

$$\operatorname{div}\frac{\mathbf{R}_0}{R_0^3} = \mathbf{R}_0\cdot\operatorname{grad}\frac{1}{R_0^3}+\frac{1}{R_0^3}\operatorname{div}\mathbf{R}_0 = 0$$

and

$$(\overline{\mathbf{m}}\cdot\nabla)\frac{\mathbf{R}_0}{R_0^3} = \frac{1}{R_0^3}(\overline{\mathbf{m}}\cdot\nabla)\mathbf{R}_0+\mathbf{R}_0(\overline{\mathbf{m}}\cdot\nabla)\frac{1}{R_0^3} = \frac{\overline{\mathbf{m}}}{R_0^3}-\frac{3\mathbf{R}_0(\overline{\mathbf{m}}\cdot\mathbf{R}_0)}{R_0^5}\,.$$

Thus,

$$\bar{\mathbf{H}} = \frac{3\mathbf{n}(\overline{\mathbf{m}}\cdot\mathbf{n})-\overline{\mathbf{m}}}{R_0^3}, \tag{66.4}$$

where $\mathbf{n}$ is again the unit vector along $\mathbf{R}_0$. We see that the magnetic field is expressed in terms of the magnetic moment by the same formula by which the electric field was expressed in terms of the dipole moment [see (62.5)].

If all the charges of the system have the same ratio of charge to mass, then we can write

$$\mathbf{m} = \frac{1}{2c}\sum e\mathbf{r}\times\mathbf{v} = \frac{e}{2mc}\sum m\mathbf{r}\times\mathbf{v}.$$

If the velocities of all the charges $v \ll c$, then $m\mathbf{v}$ is the momentum $\mathbf{p}$ of the charge and we get

$$\mathbf{m} = \frac{e}{2mc}\sum \mathbf{r}\times\mathbf{p} = \frac{e}{2mc}\mathbf{M}, \tag{66.5}$$

where $\mathbf{M} = \Sigma\mathbf{r}\times\mathbf{p}$ is the mechanical angular momentum of the system. Thus in this case, the ratio of magnetic moment to angular momentum is constant and equal to $e/2mc$.

PROBLEM

Find the ratio of the magnetic moment to the angular momentum for a system of two charges (velocities $v \ll c$).

SOLUTION. Choosing the origin of coordinates at the centre of mass of the two particles we have $m_1\mathbf{r}_1+m_2\mathbf{r}_2 = 0$ and $\mathbf{p}_1 = -\mathbf{p}_2 = \mathbf{p}$, where $\mathbf{p}$ is the momentum of the relative motion. With the aid of these relations, we find

$$\mathbf{m} = \frac{1}{2c}\left(\frac{e_1}{m_1^2}+\frac{e_2}{m_2^2}\right)\frac{m_1m_2}{m_1+m_2}\mathbf{M}.$$

§67. Larmor precession

Let us consider a system of charges in an external constant uniform magnetic field.

The time average of the force acting on the system,

$$\overline{\mathbf{F}} = \sum\frac{e}{c}\overline{\mathbf{v}\times\mathbf{H}} = \frac{d}{dt}\overline{\sum\frac{e}{c}\mathbf{r}\times\mathbf{H}},$$

is zero, as is the time average of the time derivative of any quantity which varies over a finite range. The average value of the moment of the forces is

$$\overline{\mathbf{K}} = \sum \frac{e}{c} \overline{(\mathbf{r} \times (\mathbf{v} \times \mathbf{H}))}$$

and is different from zero. It can be expressed in terms of the magnetic moment of the system, by expanding the vector triple product:

$$\mathbf{K} = \sum \frac{e}{c} \{\mathbf{v}(\mathbf{r} \cdot \mathbf{H}) - \mathbf{H}(\mathbf{v} \cdot \mathbf{r})\} = \sum \frac{e}{c} \left\{\mathbf{v}(\mathbf{r} \cdot \mathbf{H}) - \tfrac{1}{2} \mathbf{H} \frac{\mathrm{d}}{\mathrm{d}t} \mathbf{r}^2 \right\}.$$

The second term gives zero after averaging, so that

$$\overline{\mathbf{K}} = \sum \frac{e}{c} \overline{\mathbf{v}(\mathbf{r} \cdot \mathbf{H})} = \frac{1}{2c} \sum e \{\overline{\mathbf{v}(\mathbf{r} \cdot \mathbf{H})} - \overline{\mathbf{r}(\mathbf{v} \cdot \mathbf{H})}\}$$

[the last transformation is analogous to the one used in deriving (66.3)], or finally

$$\overline{\mathbf{K}} = \overline{\mathbf{m}} \times \mathbf{H}. \tag{67.1}$$

We call attention to the analogy with formula (64.6) for the electrical case.

We now consider a system of identical charges executing a finite motion (with velocities $v \ll c$) in the centrally symmetric electric field produced by a certain fixed charge (say, the system of electrons of an atom in the field of the nucleus). We assume that this system is in a weak uniform magnetic field.

In the absence of the external field the total angular momentum **M** of the system would be a constant. The presence of the weak magnetic field results in a slow change of **M** with time. Let us consider the character of this change. To eliminate the effect of the rapidly varying intrinsic motion of the charges in the system, we average **M** over the period of this motion.

According to the familiar equation of mechanics [cf. (27.3)],

$$\mathrm{d}\overline{\mathbf{M}}/\mathrm{d}t = \overline{\mathbf{K}},$$

where $\overline{\mathbf{K}}$ is the torque of the external forces acting on the system (averaged over the same time interval as was **M**). According to (67.1) and (66.5) we have

$$\overline{\mathbf{K}} = \overline{\mathbf{m}} \times \mathbf{H} = \frac{e}{2mc} \overline{\mathbf{M}} \times \mathbf{H}.$$

Therefore

$$\mathrm{d}\overline{\mathbf{M}}/\mathrm{d}t = -\boldsymbol{\Omega} \times \overline{\mathbf{M}}, \qquad (67.2)$$

where

$$\boldsymbol{\Omega} = e\mathbf{H}/2mc. \qquad (67.3)$$

An equation of the form of (67.2) means that the vector $\overline{\mathbf{M}}$ (and with it the magnetic moment $\overline{\mathbf{m}}$) rotates with angular velocity $-\boldsymbol{\Omega}$ around the direction of the field, keeping fixed its absolute magnitude and the angle it makes with the field direction. This phenomenon is called the *Larmor precession*, and the angular velocity (67.3) the *Larmor frequency*.

We can now make more precise what we meant above by a sufficiently weak field: the Larmor frequency Ω must be small compared with the frequencies of the intrinsic finite motion of the charges of the system. It is obvious that only then is it meaningful to consider the change with time of the angular momentum averaged in the fashion described above.

CHAPTER 13

ELECTROMAGNETIC WAVES

§68. The wave equation

The electromagnetic field in vacuum is determined by Maxwell's equations, in which we must put $\varrho = 0$, $\mathbf{j} = 0$. We write them once more:

$$\operatorname{curl} \mathbf{E} = -\frac{1}{c}\frac{\partial \mathbf{H}}{\partial t}, \quad \operatorname{div} \mathbf{H} = 0, \tag{68.1}$$

$$\operatorname{curl} \mathbf{H} = \frac{1}{c}\frac{\partial \mathbf{E}}{\partial t}, \quad \operatorname{div} \mathbf{E} = 0. \tag{68.2}$$

These equations may possess non-zero solutions. This means that an electromagnetic field can exist even in the absence of any charges.

Electromagnetic fields occurring in vacuum in the absence of charges are called *electromagnetic waves*. We now take up the study of the properties of such fields.

First of all we note that such fields must necessarily be time-varying. For, in the contrary case, $\partial \mathbf{H}/\partial t = \partial \mathbf{E}/\partial t = 0$ and the equations (68.1) and (68.2) go over into the equations (59.1), (59.2) and (65.1), (65.2) of a constant field, in which, however, we now have $\varrho = 0$, $\mathbf{j} = 0$. But the solutions of these equations which are given by formulae (59.9) and (65.5) become zero for $\varrho = 0$, $\mathbf{j} = 0$.

Let us derive the equations determining the potentials of electromagnetic waves.

As we already know, because of the ambiguity in the potentials we can always subject them to an auxiliary condition. For this reason, we choose the potentials of the electromagnetic waves so that the scalar potential is zero:

$$\phi = 0. \tag{68.3}$$

Then

$$\mathbf{E} = -\frac{1}{c}\frac{\partial \mathbf{A}}{\partial t}, \qquad \mathbf{H} = \operatorname{curl} \mathbf{A}. \tag{68.4}$$

Substituting these two expressions in the first equation (68.2), we get

$$\operatorname{curl}\operatorname{curl} \mathbf{A} = -\triangle \mathbf{A} + \operatorname{grad}\operatorname{div} \mathbf{A} = -\frac{1}{c^2}\frac{\partial^2 \mathbf{A}}{\partial t^2}. \tag{68.5}$$

Despite the fact that we have already imposed one auxiliary condition on the potentials, the potential **A** is still not completely unique: we can add to it the gradient of an arbitrary function which does not depend on the time (meanwhile leaving ϕ unchanged). In particular, we can choose the potential of the electromagnetic wave so that

$$\operatorname{div} \mathbf{A} = 0. \tag{68.6}$$

For, substituting **E** from (68.4) in div **E** = 0, we have

$$\operatorname{div}\frac{\partial \mathbf{A}}{\partial t} = \frac{\partial}{\partial t}\operatorname{div} \mathbf{A} = 0,$$

that is, div **A** is a function only of the coordinates. This function can always be made zero by adding to **A** the gradient of a suitable time-independent function.

The equation (68.5) now becomes

$$\triangle \mathbf{A} - \frac{1}{c^2}\frac{\partial^2 \mathbf{A}}{\partial t^2} = 0. \tag{68.7}$$

This is the equation which determines the potential of electromagnetic waves. It is called *d'Alembert's equation*, or the *wave equation*.

Applying to (68.7) the operators curl and $\partial/\partial t$, we can verify that the electric and magnetic fields **E** and **H** satisfy the same wave equation.

§69. Plane waves

Let us consider the special case of electromagnetic waves in which the field depends only on one coordinate, say x (and on the time). Such waves are said to be *plane*. In this case the equation for the field becomes

$$\frac{\partial^2 f}{\partial t^2} - c^2 \frac{\partial^2 f}{\partial x^2} = 0, \tag{69.1}$$

where by f is understood any component of the vectors $\mathbf{E}$ and $\mathbf{H}$.

To solve this equation, we rewrite it in the form

$$\left(\frac{\partial}{\partial t} - c\frac{\partial}{\partial x}\right)\left(\frac{\partial}{\partial t} + c\frac{\partial}{\partial x}\right) f = 0,$$

and introduce new variables

$$\xi = t - \frac{x}{c}, \quad \eta = t + \frac{x}{c}$$

so that $t = \frac{1}{2}(\eta+\xi)$, $x = \frac{1}{2}c(\eta-\xi)$. Then

$$\frac{\partial}{\partial \xi} = \frac{1}{2}\left(\frac{\partial}{\partial t} - c\frac{\partial}{\partial x}\right), \quad \frac{\partial}{\partial \eta} = \frac{1}{2}\left(\frac{\partial}{\partial t} + c\frac{\partial}{\partial x}\right),$$

so that the equation for f becomes

$$\frac{\partial^2 f}{\partial \xi\, \partial \eta} = 0.$$

The solution obviously has the form $f = f_1(\xi) + f_2(\eta)$, where f_1 and f_2 are arbitrary functions. Thus

$$f = f_1\left(t - \frac{x}{c}\right) + f_2\left(t + \frac{x}{c}\right). \tag{69.2}$$

Suppose, for example, $f_2 = 0$, so that

$$f = f_1\left(t - \frac{x}{c}\right).$$

Let us clarify the meaning of this solution. In each plane x = constant, the field changes with the time; at each given moment, the field is different for different x. It is clear that the field has the same values for coordinates x and times t which satisfy the relation $t-(x/c) =$ = constant, that is,

$$x = \text{constant}+ct.$$

This means that if, at some time $t = 0$, the field at a certain point x in space had some definite value, then after an interval of time t the field has that same value at a distance ct along the x axis from the original place. We can say that all the values of the electromagnetic field are propagated in space along the x axis with a velocity equal to the velocity of light, c.

Thus, $f_1(t-x/c)$ represents a plane wave moving in the positive direction along the x axis. It is easy to show that $f_2(t+x/c)$ represents a wave moving in the opposite, negative, direction along the x axis.

In §68 we showed that the potentials of the electromagnetic wave can be chosen so that $\phi = 0$, and div $\mathbf{A} = 0$. We choose in this same way the potentials of the plane wave which we are now considering. The condition div $\mathbf{A} = 0$ gives in this case

$$\partial A_x/\partial x = 0,$$

since all quantities are independent of y and z. According to (69.1) we then have also $\partial^2 A_x/\partial t^2 = 0$, that is, $\partial A_x/\partial t$ = constant. But the derivative $\partial \mathbf{A}/\partial t$ determines the electric field, and we see that the non-zero component A_x represents in this case the presence of a constant longitudinal electric field. Since such a field has no relation to the electromagnetic wave, we can put $A_x = 0$.

Thus the vector potential of the plane wave can always be chosen perpendicular to the x axis, i.e. to the direction of propagation of that wave.

Let us consider a plane wave moving in the positive direction of the x axis; in this wave, all quantities, and in particular $\mathbf{A}$, are functions only of $t-(x/c)$. From the formulae

$$\mathbf{E} = -\frac{1}{c}\frac{\partial \mathbf{A}}{\partial t}, \quad \mathbf{H} = \text{curl}\,\mathbf{A},$$

we therefore obtain

$$\mathbf{E} = -\frac{1}{c}\mathbf{A}', \quad \mathbf{H} = \nabla\times\mathbf{A} = \nabla\left(t-\frac{x}{c}\right)\times\mathbf{A}' = -\frac{1}{c}\mathbf{n}\times\mathbf{A}', \tag{69.3}$$

where the prime denotes differentiation with respect to $t-(x/c)$ and $\mathbf{n}$ is a unit vector along the direction of propagation of the wave. Substituting the first equation in the second, we obtain

$$\mathbf{H} = \mathbf{n}\times\mathbf{E}. \tag{69.4}$$

We see that the electric and magnetic fields $\mathbf{E}$ and $\mathbf{H}$ of a plane wave are directed perpendicular to the direction of propagation of the wave. For this reason, electromagnetic waves are said to be *transverse*. From (69.4) it is clear also that the electric and magnetic fields of the plane wave are perpendicular to each other and equal to each other in absolute value.

The energy flux in the plane wave is

$$\mathbf{S} = \frac{c}{4\pi}\mathbf{E}\times\mathbf{H} = \frac{c}{4\pi}E^2\mathbf{n} = \frac{c}{4\pi}H^2\mathbf{n}.$$

Thus the energy flux is directed along the direction of propagation of the wave. Since

$$W = \frac{1}{8\pi}(E^2+H^2) = \frac{E^2}{4\pi}$$

is the energy density of the wave, we can write

$$\mathbf{S} = cW\mathbf{n}, \tag{69.5}$$

in accordance with the fact that the field propagates with the velocity of light.

The momentum per unit volume of the electromagnetic field is $\mathbf{S}/c^2$. For a plane wave this gives $(W/c)\mathbf{n}$. We call attention to the fact that the relation between the energy W and the momentum W/c for the electromagnetic wave is the same as for a particle moving with the velocity of light [see (39.12)].

The flux of momentum of the field is determined by the components σ_{ik} of the Maxwell stress tensor (58.5). Again choosing the direction of propagation of the wave as the x axis, we find that the only non-zero component is

$$\sigma_{xx} = W. \tag{69.6}$$

As it must be, the flux of momentum is along the direction of propagation of the wave, and is equal in magnitude to the energy density.

PROBLEM

Determine the force exerted on a wall from which an incident plane electromagnetic wave is reflected (with reflection coefficient R).

SOLUTION. The force $\mathbf{f}$ acting on unit area of the wall is given by the flux of momentum through this area, i.e., it is the vector with components

$$f_i = \sigma_{ik}N_k + \sigma'_{ik}N_k$$

where $\mathbf{N}$ is the vector normal to the surface of the wall, and σ_{ik} and σ'_{ik} are the components of the stress tensors for the incident and reflected waves. Using (69.6), we obtain

$$\mathbf{f} = W\mathbf{n}(\mathbf{N}\cdot\mathbf{n}) + W'\mathbf{n}'(\mathbf{N}\cdot\mathbf{n}').$$

From the definition of the reflection coefficient, we have $W' = RW$. Also introducing the angle of incidence θ (which is equal to the reflection angle) and writing out components, we find the normal force (*light pressure*)

$$f_N = W(1+R)\cos^2\theta$$

and the tangential force

$$f_t = W(1-R)\sin\theta\cos\theta.$$

§70. Monochromatic plane waves

A very important special case of electromagnetic waves is a wave in which the field is a simply periodic function of the time. Such a wave is said to be *monochromatic*. All quantities (potentials, field components) in a monochromatic wave depend on the time through a factor of the form $\cos(\omega t+\alpha)$. The quantity ω is called the *angular frequency* of the wave (we shall simply call it the *frequency*).

In a plane wave (propagating along the x axis), the field is a function only of $t-(x/c)$. Therefore, if the plane wave is monochromatic, its field is a simply periodic function of $t-(x/c)$. The vector potential of such a wave is most conveniently written as the real part of a complex expression:

$$\mathbf{A} = \mathrm{Re}\,\{\mathbf{A}_0 e^{-i\omega(t-x/c)}\}. \tag{70.1}$$

Here $\mathbf{A}_0$ is a certain constant complex vector. Obviously, the fields **E** and **H** of such a wave have analogous forms with the same frequency ω. The quantity

$$\lambda = 2\pi c/\omega \tag{70.2}$$

is called the *wavelength*; it is the period of variation of the field with the coordinate x at a fixed time t.

The vector

$$\mathbf{k} = \frac{\omega}{c}\mathbf{n} \tag{70.3}$$

(where **n** is a unit vector along the direction of propagation of the wave) is called the *wave vector*. In terms of it we can write (70.1) in the form

$$\mathbf{A} = \mathrm{Re}\,\{\mathbf{A}_0 e^{i(\mathbf{k}\cdot\mathbf{r}-\omega t)}\}, \tag{70.4}$$

which is independent of the choice of coordinate axes. The quantity which appears multiplied by i in the exponent is called the *phase* of the wave.

So long as we perform only linear operations, we can omit the sign Re for taking the real part, and operate with complex quantities as such.† Thus, substituting

$$\mathbf{A} = \mathbf{A}_0 e^{i(\mathbf{k}\cdot\mathbf{r}-\omega t)}$$

---

† If two quantities $\mathbf{A}(t)$ and $\mathbf{B}(t)$ are written in complex form

$$\mathbf{A}(t) = \mathbf{A}_0 e^{-i\omega t}, \qquad \mathbf{B}(t) = \mathbf{B}_0 e^{-i\omega t},$$

then in forming their product we must first, of course, separate out the real part. But if, as frequently happens, we are interested only in the time average of this product, it can be computed as

$$\tfrac{1}{2}\,\mathrm{Re}\,\{\mathbf{A}\cdot\mathbf{B}^*\},$$

since we have

$$\mathrm{Re}\,\mathbf{A}\cdot\mathrm{Re}\,\mathbf{B} = \tfrac{1}{4}(\mathbf{A}_0 e^{-i\omega t}+\mathbf{A}_0^* e^{i\omega t})\cdot(\mathbf{B}_0 e^{-i\omega t}+\mathbf{B}_0^* e^{i\omega t}),$$

and when we average, the terms containing factors $e^{\pm 2i\omega t}$ vanish, so that we are left with

$$\overline{\mathrm{Re}\,\mathbf{A}\cdot\mathrm{Re}\,\mathbf{B}} = \tfrac{1}{4}(\mathbf{A}\cdot\mathbf{B}^*+\mathbf{A}^*\cdot\mathbf{B}) = \tfrac{1}{2}\,\mathrm{Re}\,(\mathbf{A}\cdot\mathbf{B}^*).$$

in (69.3), we find the relation between the intensities and the vector potential of a plane monochromatic wave in the form

$$\mathbf{E} = ik\mathbf{A}, \quad \mathbf{H} = i\mathbf{k}\times\mathbf{A}. \tag{70.5}$$

We shall now treat in more detail the direction of the field of a monochromatic wave. To be specific, we shall talk of the electric field

$$\mathbf{E} = \mathrm{Re}\{\mathbf{E}_0\, e^{i(\mathbf{k}\cdot\mathbf{r}-\omega t)}\}$$

(everything stated below applies equally well, of course, to the magnetic field). The quantity $\mathbf{E}_0$ is a certain complex vector. Its square $\mathbf{E}_0^2$ is (in general) a complex number. If the argument of this number is -2α (i.e. $\mathbf{E}_0^2 = |\mathbf{E}_0^2|e^{-2i\alpha}$), the vector $\mathbf{b}$ defined by

$$\mathbf{E}_0 = \mathbf{b}e^{-i\alpha} \tag{70.6}$$

will have its square real, $\mathbf{b}^2 = |\mathbf{E}_0|^2$. With this definition, we write

$$\mathbf{E} = \mathrm{Re}\{\mathbf{b}e^{i(\mathbf{k}\cdot\mathbf{r}-\omega t-\alpha)}\}. \tag{70.7}$$

We write $\mathbf{b}$ in the form

$$\mathbf{b} = \mathbf{b}_1 + i\mathbf{b}_2,$$

where $\mathbf{b}_1$ and $\mathbf{b}_2$ are real vectors. Since $\mathbf{b}^2 = \mathbf{b}_1^2 - \mathbf{b}_2^2 + 2i\mathbf{b}_1\cdot\mathbf{b}_2$ must be a real quantity, $\mathbf{b}_1\cdot\mathbf{b}_2 = 0$, i.e. the vectors $\mathbf{b}_1$ and $\mathbf{b}_2$ are mutually perpendicular. We choose the direction of $\mathbf{b}_1$ as the y axis (and the x axis along the direction of propagation of the wave). We then have from (70.7)

$$\left.\begin{aligned} E_y &= b_1 \cos(\omega t - \mathbf{k}\cdot\mathbf{r}+\alpha), \\ E_z &= \pm b_2 \sin(\omega t - \mathbf{k}\cdot\mathbf{r}+\alpha), \end{aligned}\right\} \tag{70.8}$$

where we use the plus (minus) sign if $\mathbf{b}_2$ is along the positive (negative) z axis. From (70.8) it follows that

$$\frac{E_y^2}{b_1^2} + \frac{E_z^2}{b_2^2} = 1. \tag{70.9}$$

Thus we see that, at each point in space, the electric field vector rotates in a plane perpendicular to the direction of propagation of the wave, while its endpoint describes the ellipse (70.9). Such a wave is said to be *elliptically polarised.* The rotation occurs in the direction of (opposite to) a right-handed screw rotating along the x axis, if we have the plus (minus) sign in (70.8).

If $b_1 = b_2$, the ellipse (70.9) reduces to a circle, i.e. the vector $\mathbf{E}$ rotates while remaining constant in magnitude. In this case we say that the wave is *circularly polarised.* The choice of the directions of the y and z axes is now obviously arbitrary. We note that in such a wave the ratio of the y and z components of the complex amplitude $\mathbf{E}_0$ is

$$E_{0z}/E_{0y} = \pm i \tag{70.10}$$

for rotation in the same (opposite) direction as that of a right-handed screw (*right* and *left* polarisations).†

Finally, if b_1 or b_2 equals zero, the field of the wave is everywhere and always parallel (or antiparallel) to one and the same direction. In this case the wave is said to be *linearly polarised,* or plane polarised. An elliptically polarised wave can clearly be treated as the superposition of two linearly polarised waves.

§71. The Doppler effect

Now let us turn to the definition of the wave vector and introduce the four-dimensional wave vector with components

$$k^\mu = (\omega/c,\ \mathbf{k}). \tag{71.1}$$

That these quantities actually form a four-vector is obvious from the fact that we get a scalar (the phase of the wave) when we multiply by x^μ:

$$k_\mu x^\mu = \omega t - \mathbf{k}\cdot\mathbf{r}. \tag{71.2}$$

From the definitions (70.3) and (71.1) we see that the square of the wave four-vector is zero:

$$k^\mu k_\mu = 0. \tag{71.3}$$

† We assume that the coordinate axes form a right-handed system.

Using the law of transformation of the wave four-vector we can easily treat the *Doppler effect*—the change in frequency ω of the wave emitted by a source moving with respect to the observer, as compared to the "true" frequency ω_0 of the same source in the reference frame (K_0) in which it is at rest.

Let V be the velocity of the source, i.e. the velocity of the frame K_0 relative to K. According to the general formula for transformation of four-vectors, we have

$$k^{(0)0} = \frac{k^0-(Vk^1/c)}{\sqrt{[1-(V^2/c^2)]}}$$

(the velocity of the frame K relative to K_0 is $-V$). Substituting $k^0 = \omega/c$, $k^1 = k \cos \alpha = (\omega/c) \cos \alpha$, where α is the angle (in the frame K) between the direction of emission of the wave and the direction of motion of the source, and expressing ω in terms of ω_0, we obtain

$$\omega = \omega_0 \frac{\sqrt{[1-(V^2/c^2)]}}{1-(V/c)\cos\alpha}. \qquad (71.4)$$

This is the required formula. For $V \ll c$, and if the angle α is not too close to $\pi/2$, it gives

$$\omega \approx \omega_0\left(1+\frac{V}{c}\cos\alpha\right). \qquad (71.5)$$

For $\alpha = \pi/2$, we have

$$\omega = \omega_0\sqrt{[1-(V^2/c^2)]} \approx \omega_0[1-(V^2/2c^2)]; \qquad (71.6)$$

in this case the relative change in frequency is proportional to the square of the ratio V/c.

§72. Spectral resolution

Every wave can be subjected to the process of *spectral resolution*, i.e. can be represented as a superposition of monochromatic waves with various frequencies. The character of this expansion varies according to the character of the time dependence of the field.

One category consists of those cases where the expansion contains frequencies forming a discrete sequence of values. The simplest case of this type arises in the resolution of a purely periodic (though not monochromatic) field. This is the usual expansion in Fourier series; it contains the frequencies which are integral multiples of the "fundamental" frequency $\omega_0 = 2\pi/T$, where T is the period of the field. We write it in the form

$$f = \sum_{n=-\infty}^{\infty} f_n e^{-i\omega_0 nt} \tag{72.1}$$

(where f is any of the quantities describing the field). The quantities f_n are defined in terms of the function f by the integrals

$$f_n = \frac{1}{T} \int_{-T/2}^{T/2} f(t) e^{in\omega_0 t}\, dt. \tag{72.2}$$

Because $f(t)$ must be real,

$$f_{-n} = f_n^*. \tag{72.3}$$

In more complicated cases, the expansion may contain integral multiples (and sums of integral multiples) of several different incommensurable fundamental frequencies.

When the sum (72.1) is squared and averaged over the time, the products of terms with different frequencies give zero because they contain oscillating factors. Only terms of the form $f_n f_{-n} = |f_n|^2$ remain. Thus the average of the square of the field, i.e. the average intensity of the wave, is the sum of the intensities of its monochromatic components:

$$\overline{f^2} = \sum_{n=-\infty}^{\infty} |f_n|^2 = 2\sum_{n=1}^{\infty} |f_n|^2 \tag{72.4}$$

(where it is assumed that the average of the function f over a period is zero, i.e. $f_0 = \bar{f} = 0$).

Another category consists of fields which are expandable in a Fourier integral containing a continuous sequence of different frequencies. For this to be possible, the function $f(t)$ must satisfy certain definite conditions; usually we consider functions which vanish for $t \to \pm\infty$. Such an expansion has the form

$$f(t) = \int_{-\infty}^{\infty} f_\omega e^{-i\omega t} \frac{d\omega}{2\pi}, \tag{72.5}$$

where the Fourier components are given in terms of the function $f(t)$ by the integrals

$$f_\omega = \int_{-\infty}^{\infty} f(t) e^{i\omega t}\, dt. \tag{72.6}$$

Analogously to (72.3),

$$f_{-\omega} = f_\omega^*. \tag{72.7}$$

Let us calculate the integral of f^2 over all time. Using (72.5) and (72.6), we have

$$\int_{-\infty}^{\infty} f^2\, dt = \int_{-\infty}^{\infty} \left\{ f \int_{-\infty}^{\infty} f_\omega e^{-i\omega t} \frac{d\omega}{2\pi} \right\} dt =$$

$$= \int_{-\infty}^{\infty} \left\{ f_\omega \int_{-\infty}^{\infty} f e^{-i\omega t}\, dt \right\} \frac{d\omega}{2\pi} = \int_{-\infty}^{\infty} f_\omega f_{-\omega} \frac{d\omega}{2\pi},$$

or, using (72.7),

$$\int_{-\infty}^{\infty} f^2\, dt = \int_{-\infty}^{\infty} |f_\omega|^2 \frac{d\omega}{2\pi} = 2\int_{0}^{\infty} |f_\omega|^2 \frac{d\omega}{2\pi}. \tag{72.8}$$

Thus the integrated intensity can be expressed in terms of the intensities of the Fourier components of the wave.

§73. Partially polarised light

Every monochromatic wave is, by definition, necessarily polarised. However, we usually have to deal with waves which are only approximately monochromatic, and which contain frequencies in a small interval $\Delta\omega$. We consider such a wave, and let ω be some average frequency for it. Then its field (to be specific we shall consider the electric field **E**) at a fixed point in space can be written in the form

$$\mathbf{E} = \mathbf{E}_0(t)e^{-i\omega t},$$

where the complex amplitude $\mathbf{E}_0(t)$ is some slowly varying function of the time (for a strictly monochromatic wave $\mathbf{E}_0$ would be constant). Since $\mathbf{E}_0$ determines the polarisation of the wave, this means that at each point of the wave its polarisation changes with time; such a wave is said to be *partially polarised.*

The polarisation properties of electromagnetic waves, and of light in particular, are observed experimentally by passing the light to be investigated through various bodies† and then observing the intensity of the transmitted light. From the mathematical point of view this means that we draw conclusions concerning the polarisation properties of the light from the values of certain quadratic functions of its field. Here of course we mean the time averages of such functions.

Quadratic functions of the field are made up of terms proportional to the products E_iE_k, $E_i^*E_k^*$ or $E_iE_k^*$. Products of the form

$$E_iE_k = E_{0i}E_{0k}e^{-2i\omega t}, \qquad E_i^*E_k^* = E_{0i}^*E_{0k}^*e^{2i\omega t},$$

which contain the rapidly oscillating factors $e^{\pm 2i\omega t}$, give zero when the time average is taken. The products $E_iE_k^* = E_{0i}E_{0k}^*$ do not contain such factors, and so their averages are not zero. Thus we see that the polarisation properties of the light are completely characterised by the tensor

$$J_{ik} = \overline{E_{0i}E_{0k}^*}. \tag{73.1}$$

Since the vector $\mathbf{E}_0$ always lies in a plane perpendicular to the direction of the wave, the tensor J_{ik} has only four components (in this section the indices i, k are understood to take on only two values: $i, k = 1, 2$, corresponding to the y and z axes; the x axis is along the direction of propagation of the wave).

---

† For example, through a Nicol prism.

The sum of the diagonal elements of the tensor J_{ik} (we denote it by J) is a real quantity—the average value of the square modulus of the vector $\mathbf{E}_0$:

$$J \equiv J_{ii} = \overline{\mathbf{E}_0 \cdot \mathbf{E}_0^*}. \tag{73.2}$$

This quantity determines the intensity of the wave, as measured by the energy flux density. To eliminate this quantity which is not directly related to the polarisation properties, we introduce in place of J_{ik} the tensor

$$\varrho_{ik} = J_{ik}/J, \tag{73.3}$$

for which $\varrho_{ii} = 1$; we call it the *polarisation tensor*.

From the definition (73.1) we see that the components of the tensor J_{ik}, and consequently also ϱ_{ik}, are related by

$$\varrho_{ik} = \varrho_{ki}^* \tag{73.4}$$

(i.e. the tensor is *Hermitian*). Consequently the diagonal components ϱ_{11} and ϱ_{22} are real (with $\varrho_{11}+\varrho_{22} = 1$) while $\varrho_{21} = \varrho_{12}^*$. Thus the polarisation tensor is characterised by three real parameters.

Let us study the conditions that the tensor ϱ_{ik} must satisfy for completely polarised light. In this case $\mathbf{E}_0$ = constant, and so we have simply

$$J_{ik} = J\varrho_{ik} = E_{0i}E_{0k}^* \tag{73.5}$$

(without averaging), i.e. the components of the tensor can be written as products of components of some constant vector. The necessary and sufficient condition for this is that the determinant vanish:

$$|\varrho_{ik}| = \varrho_{11}\varrho_{22} - \varrho_{12}\varrho_{21} = 0. \tag{73.6}$$

The opposite case is that of unpolarised or *natural* light. Complete absence of polarisation means that all directions (in the yz plane) are equivalent. In other words, the polarisation tensor must have the form

$$\varrho_{ik} = \tfrac{1}{2}\delta_{ik}. \tag{73.7}$$

The determinant is $|\varrho_{ik}| = \frac{1}{4}$.

An arbitrary tensor ϱ_{ik} can be resolved into two parts, one symmetric (in the indices i, k) and the other antisymmetric. Let us consider the special case where the latter part is absent. Because of (73.4) the symmetric tensor ϱ_{ik} is then real ($\varrho_{ik} = \varrho_{ik}^*$). Like every symmetric tensor it can be brought to principal axes, with two different eigenvalues, which we denote by λ_1 and λ_2. The directions of the principal axes are mutually perpendicular. Denoting the unit vectors along these directions by $\mathbf{n}^{(1)}$ and $\mathbf{n}^{(2)}$, we can represent ϱ_{ik} in the form

$$\varrho_{ik} = \lambda_1 n_i^{(1)} n_k^{(1)} + \lambda_2 n_i^{(2)} n_k^{(2)}, \quad \lambda_1 + \lambda_2 = 1. \tag{73.8}$$

The quantities λ_1 and λ_2 are positive and take on values from 0 to 1.

Each of the two terms in (73.8) has the form of a product of two components of a constant real vector ($\sqrt{\lambda_1}\,\mathbf{n}^{(1)}$ or $\sqrt{\lambda_2}\,\mathbf{n}^{(2)}$). In other words, each of the terms corresponds to linearly polarised light. We also see that there is no term in (73.8) containg products of components of the two waves. This means that the two parts can be regarded as physically independent, or *incoherent*. For, if two waves are independent of one another, the average value of the product $E_i^{(1)} E_k^{(2)}$ is equal to the product of the average values of the two factors, and since each of them is zero,

$$\overline{E_i^{(1)} E_k^{(2)}} = 0.$$

Thus in the case considered here the partially polarised wave can be represented as a superposition of two incoherent waves (with intensities proportional to λ_1 and λ_2), linearly polarised along mutually perpendicular directions. (In the general case of a complex tensor ϱ_{ik}, one can show that the light can be represented as a superposition of two incoherent elliptically polarised waves, whose polarisation ellipses are similar and mutually perpendicular.)

§74. Geometrical optics

A plane wave is characterised by the property that its direction of propagation and amplitude are the same everywhere. Arbitrary electromagnetic waves, of course, do not have this property.

Nevertheless, a great many electromagnetic waves, which are not plane, have the property that within each small region of space they can be considered to be plane. For this, it is clearly necessary that the amplitude and direction of the wave remain practically constant over distances of the order of the wavelength.

If this condition is satisfied, we can introduce the *wave surfaces*, i.e. surfaces at all of whose points the phase of the wave is the same (at a given time). (The wave surfaces of a plane wave are planes perpendicular to the direction of propagation of the wave.) In each small region of space we can speak of a direction of propagation of the wave, normal to the wave surface. In this way we can introduce the concept of *rays*—curves whose tangents at each point coincide with the direction of propagation of the wave.

The study of the laws of propagation of waves in this case constitutes the domain of *geometrical optics*. Consequently, geometrical optics considers the propagation of electromagnetic waves, in particular of light, as the propagation of rays, completely divorced from their wave properties. In other words, geometrical optics corresponds to the limiting case of small wavelength, $\lambda \to 0$.

We now take up the derivation of the fundamental equation of geometrical optics—the equation determining the direction of the rays. Let f be any quantity describing the field of the wave (any component of $\mathbf{E}$ or $\mathbf{H}$). For a plane monochromatic wave, f has the form

$$f = ae^{i(\mathbf{k}\cdot\mathbf{r}-\omega t+\alpha)} \tag{74.1}$$

(we omit the Re; it is understood that we take the real part of all expressions).

We write the expression for the field in the form

$$f = ae^{i\psi}. \tag{74.2}$$

If the wave is not plane, but geometrical optics is applicable, the amplitude a is, generally speaking, a function of the coordinates and time, and the phase ψ, which is called the *eikonal*, does not have a simple form, as in (74.1). It is important to note, however, that ψ is a large quantity. This is clear immediately from the fact that it changes by 2π when we move through one wavelength, and geometrical optics corresponds to the limit $\lambda \to 0$.

Over small space regions and time intervals the eikonal ψ can be expanded in series; to terms of first order, we have

$$\psi = \psi_0 + \mathbf{r}\cdot\frac{\partial\psi}{\partial\mathbf{r}} + t\,\frac{\partial\psi}{\partial t}$$

(the origin for coordinates and time has been chosen within the space region and time interval under consideration; the derivatives are evaluated at the origin). Comparing this expression with (74.1), we can write

$$\mathbf{k} = \frac{\partial\psi}{\partial\mathbf{r}} \equiv \operatorname{grad}\psi, \qquad \omega = -\frac{\partial\psi}{\partial t}, \tag{74.3}$$

which corresponds to the fact that in each small region of space (and each small interval of time) the wave can be considered as plane. From the definition of the wave vector, we have $\mathbf{k}^2 = \omega^2/c^2$. Substituting $\mathbf{k}$ and ω from (74.3), we find

$$(\nabla\psi)^2 = \frac{1}{c^2}\left(\frac{\partial\psi}{\partial t}\right)^2. \tag{74.4}$$

This first-order partial differential equation is called the *eikonal equation* and is the fundamental equation of geometrical optics.

Equation (74.4) can also be derived by a direct transition to the limit $\lambda \to 0$ in the wave equation. The field f satisfies the wave equation

$$\triangle f = \frac{1}{c^2}\frac{\partial^2 f}{\partial t^2}. \tag{74.5}$$

For a function of the form (74.2) we have

$$\frac{\partial^2 f}{\partial t^2} = \frac{\partial^2 a}{\partial t^2}e^{i\psi} + 2i\,\frac{\partial a}{\partial t}\,\frac{\partial\psi}{\partial t}\,e^{i\psi} + if\frac{\partial^2\psi}{\partial t^2} - \left(\frac{\partial\psi}{\partial t}\right)^2 f.$$

But the eikonal ψ in geometrical optics is a large quantity. We can therefore neglect the first three terms compared with the fourth, so that

$$\frac{\partial^2 f}{\partial t^2} \approx -\left(\frac{\partial \psi}{\partial t}\right)^2 f.$$

Similarly, we find

$$\triangle f \approx -(\nabla \psi)^2 f,$$

and substitution in (74.5) gives (74.4).

From the form of the eikonal equation there results a remarkable analogy between geometrical optics and the mechanics of material particles. The motion of a material particle is determined by the Hamilton–Jacobi equation for the action S (§31). This equation, like the eikonal equation, is an equation in first partial derivatives. The action S is related to the momentum $\mathbf{p}$ and the Hamiltonian $\mathscr{H}$ of the particle by the relations

$$\mathbf{p} = \frac{\partial S}{\partial \mathbf{r}}, \quad \mathscr{H} = -\frac{\partial S}{\partial t}.$$

Comparing these formulae with the formulae (74.3), we see that the wave vector plays the same role in geometrical optics as the momentum of the particle in mechanics, while the frequency plays the role of the Hamiltonian, i.e., the energy of the particle. The absolute magnitude of the wave vector is related to the frequency by the formula $k = \omega/c$. This relation is analogous to the relation $p = \mathscr{E}/c$ between the momentum and energy of a particle with zero mass, and its velocity which is equal to the velocity of light.

For a particle, we have the Hamilton equations

$$\dot{\mathbf{p}} = -\frac{\partial \mathscr{H}}{\partial \mathbf{r}}, \quad \mathbf{v} = \dot{\mathbf{r}} = \frac{\partial \mathscr{H}}{\partial \mathbf{p}}.$$

In view of the analogy we have pointed out, we can immediately write the corresponding equations for rays:

$$\dot{\mathbf{k}} = -\frac{\partial \omega}{\partial \mathbf{r}}, \quad \dot{\mathbf{r}} = \frac{\partial \omega}{\partial \mathbf{k}}. \tag{74.6}$$

In vacuum, $\omega = ck$, so that $\dot{\mathbf{k}} = 0$, $\mathbf{v} = c\mathbf{n}$ (**n** is a unit vector along the direction of propagation); in other words, as it must be, in vacuum the rays are straight lines, along which the light travels with velocity c.†

§75. The limits of geometrical optics

From the definition of a monochromatic plane wave, its amplitude is the same everywhere and at all times. Such a wave is infinite in extent in all directions in space, and exists over the whole range of time from $-\infty$ to $+\infty$. Any wave whose amplitude is not constant everywhere at all times can only be more or less monochromatic. We now take up the question of the "degree of non-monochromaticity" of a wave.

Let us consider an electromagnetic wave whose amplitude at each point is a function of the time. Let ω_0 be some average frequency of the wave. Then the field of the wave, for example the electric field, at a given point has the form $\mathbf{E}_0(t)e^{-i\omega_0 t}$. This field, although it is of course not monochromatic, can be expanded in monochromatic waves, that is, in a Fourier integral. The amplitude of the component in this expansion, with frequency ω, is proportional to the integral

$$\int_{-\infty}^{+\infty} \mathbf{E}_0(t)e^{i(\omega-\omega_0)t}\,dt.$$

The factor $e^{i(\omega-\omega_0)t}$ is a periodic function whose average value is zero. If $\mathbf{E}_0$ were exactly constant, then the integral would be exactly zero, for $\omega \neq \omega_0$. If, however, $\mathbf{E}_0(t)$ is variable, but hardly changes over a time interval of order $1/|\omega-\omega_0|$, then the integral is almost equal to zero, the more nearly the slower the variation of $\mathbf{E}_0$. In order for the integral to be significantly different from zero, it is necessary

† It is true that when applied to the propagation of light in vacuum these equations lead to results already known, but the important point is that in their general form these results are also applicable to the propagation of light in material media. In just this case we find an analogy to the motion of particles in an external force field.

that $\mathbf{E}_0(t)$ vary significantly over a time interval of the order of $1/|\omega-\omega_0|$.

We denote by Δt the order of magnitude of the time interval during which the amplitude of the wave at a given point in space changes significantly. From these considerations, it now follows that the frequencies deviating most from ω_0, which appear with appreciable intensity in the spectral resolution of this wave, are determined by the condition $1/|\omega-\omega_0| \sim \Delta t$. If we denote by $\Delta\omega$ the frequency interval (around the average frequency ω_0) which enters into the spectral resolution of the wave, then we have the relation

$$\Delta\omega\,\Delta t \sim 1. \tag{75.1}$$

We see that a wave is the more monochromatic (i.e. the smaller $\Delta\omega$) the larger Δt, i.e. the slower the variation of the amplitude at a given point in space.

Relations similar to (75.1) are easily derived for the wave vector. Let Δx, Δy, Δz be the orders of magnitude of distances along the x, y, z axes, in which the wave amplitude changes significantly. At a given time, the field of the wave as a function of the coordinates has the form

$$\mathbf{E}_0(\mathbf{r})e^{i\mathbf{k}_0\cdot\mathbf{r}},$$

where $\mathbf{k}_0$ is some average value of the wave vector. By a completely analogous derivation to that for (75.1) we can obtain the interval $\Delta\mathbf{k}$ of values contained in the expansion of the wave into a Fourier integral:

$$\Delta k_x\,\Delta x \sim 1, \quad \Delta k_y\,\Delta y \sim 1, \quad \Delta k_z\,\Delta z \sim 1. \tag{75.2}$$

Let us consider, in particular, a wave which is radiated during a finite time interval. We denote by Δt the order of magnitude of this interval. The amplitude at a given point in space changes significantly during the time Δt in the course of which the wave travels completely past the point. Because of the relation (75.1) we can now say that the "lack of monochromaticity" of such a wave, $\Delta\omega$, cannot be smaller than $1/\Delta t$ (it can of course be larger):

$$\Delta\omega \gtrsim \frac{1}{\Delta t}. \tag{75.3}$$

Similarly, if Δx, Δy, Δz are the orders of magnitude of the extension of the wave in space, then for the spread in the values of components of the wave vector, entering into the resolution of the wave, we obtain

$$\Delta k_x \gtrsim \frac{1}{\Delta x}, \quad \Delta k_y \gtrsim \frac{1}{\Delta y}, \quad \Delta k_z \gtrsim \frac{1}{\Delta z}. \tag{75.4}$$

From these formulae it follows that, if we have a beam of light of finite width, then the direction of propagation of the light in such a beam cannot be strictly constant. Taking the x axis along the (average) direction of light in the beam, we obtain

$$\theta_y \gtrsim \frac{1}{k\,\Delta y} \sim \frac{\lambda}{\Delta y}, \tag{75.5}$$

where θ_y is the order of magnitude of the deviation of the beam from its average direction in the xy plane and λ is the wavelength.

On the other hand, formula (75.5) answers the question of the limit of sharpness of optical image formation. A beam of light whose rays, according to geometrical optics, would all intersect in a point, actually gives an image not in the form of a point but in the form of a spot. For the width Δ of this spot, we obtain, according to (75.5),

$$\Delta \sim \frac{1}{k\theta} \sim \frac{\lambda}{\theta}, \tag{75.6}$$

where θ is the opening angle of the beam. This formula can be applied not only to the image but also to the object. Namely, we can state that in observing a beam of light emerging from a luminous point, this point cannot be distinguished from a body of dimensions λ/θ. In this way formula (75.6) determines the limiting *resolving power* of a microscope. The minimum value of Δ, which is reached for $\theta \sim 1$, is λ, in complete agreement with the fact that the limit of geometrical optics is determined by the wavelength of the light.

PROBLEM

Determine the order of magnitude of the smallest width of a light beam produced from a parallel beam at a distance l from a diaphragm.

SOLUTION. Denoting the size of the aperture in the diaphragm by d, we have from (75.5) that the angle of deflection of the beam (the "diffraction angle") is $\sim \lambda/d$, so that the width of the beam is of order $d+(\lambda/d)l$. The smallest value of this quantity is $\sim \sqrt{(\lambda l)}$.

§76. Fresnel diffraction

The deviations from the laws of geometrical optics due to the finite wavelength of light cause the phenomena of *diffraction*. Diffraction phenomena can be observed, for example, if along the path of propagation of the light there is an obstacle—an opaque body (we call it a *screen*) or if the light passes through holes in opaque screens. If the laws of geometrical optics were strictly satisfied, there would be beyond the screen regions of *shadow* sharply delineated from regions where light falls. The diffraction has the consequence that, instead of a sharp boundary between light and shadow, there is a quite complex distribution of the intensity of the light. These diffraction phenomena appear the more strongly the smaller the dimensions of the screens and the apertures in them, relative to the wavelength.

The problem of diffraction can be discussed in general form in cases of small deviation from geometrical optics, i.e. when firstly, the dimensions of all bodies are large compared with the wavelength (this requirement applies both to the dimensions of screens and also to the distances from them to the points of emission and observation of the light); and secondly when there are only small deviations of the light from the directions of the rays given by geometrical optics.

To be specific, let us consider a screen with an aperture through which the light passes. Figure 30 shows the screen in profile (the vertical line); the light travels from left to right. We denote by u some one of the components of **E** or **H**, a function only of the coordinates, i.e. without the factor $e^{-i\omega t}$. Our problem is to determine the field u_P (and hence the light intensity $\sim |u_P|^2$) at any point of observation P beyond the screen. For an approximate solution of this problem in cases where the deviations from geometrical optics are small, we may assume that at the points of the aperture the field is the same as it would have been in the absence of the screen.

We introduce some surface S which covers the aperture in the screen and is bounded by its edges (a profile of such a surface is shown in Fig. 30 as a dashed line). We break up this surface into sections with area df, whose dimensions are small compared with the size of the aperture, but large compared with the wavelength of the light.

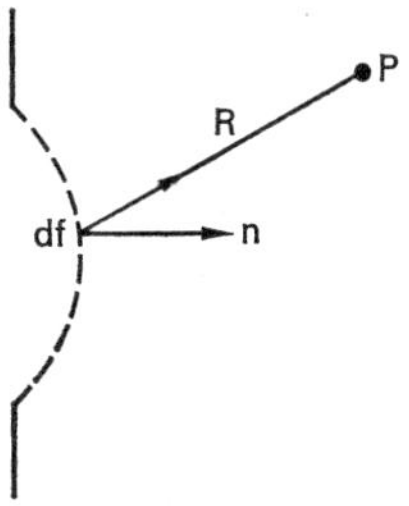

FIG. 30

We can then consider each of these sections through which the light passes as if it were itself a source of light waves spreading out on all sides from this section. We shall consider the field at the point P to be the result of superposition of the fields of waves coming from all the sections of the surface S. (This is called *Huygens' principle.*)

The field produced at the point P by the section df is proportional to the value u of the field at the section df itself. In addition, it is proportional to the projection df_n of the area df on the plane perpendicular to the direction $\mathbf{n}$ of the ray coming from the light source. This follows from the fact that no matter what position the element df has, the same rays will pass through it provided its projection df_n remains fixed, and therefore its effect on the field at P will be the same.

When the wave is propagated from df to P, its phase changes by kR (where k is the magnitude of the wave vector of the light and R is the distance from df to P); this leads to a factor $\exp(ikR)$. The wave amplitude decreases over this distance by a factor† $1/R$. Thus the contribution to the field u_P from each element of the surface S is proportional to $\mathrm{d}f_n \cdot u e^{ikR}/R$.

The total value of u_P is found by integrating over S:

$$u_P = \text{constant} \times \int u \frac{e^{ikR}}{R} \mathrm{d}f_n; \tag{76.1}$$

this is the mathematical form of Huygens' principle. In the approximation we are considering, the field u_P depends only on the shape of the edge of the aperture, not on that of the screen itself or on the material of which it is made.

If a light source Q and a point of observation P are at finite distances from the screen, only a small part of the surface S near its intersection with the line QP will contribute significantly to the integral (76.1): since the deviations from geometrical optics are small, the intensity of the light arriving at P from various points of the surface S decreases very rapidly as we move away from this line, which corresponds to propagation along a geometrical ray from source to observation point. Diffraction phenomena in which only a small portion of the wave surface plays a role are called *Fresnel diffraction* phenomena.

Let us consider the Fresnel diffraction by the straight edge of a screen. We choose the xz plane so that it passes through the point Q and the point of observation P, and the y axis on the line of the edge of the screen, as in Fig. 31.

Let the distance from the light source Q to the origin be D_q. We denote the x coordinate of the point of observation P by D_p, and its z coordinate, i.e. its distance from the xy plane, by d; the region $d < 0$ below the xy plane is the region which according to geometrical optics should be in shadow (region of geometrical shadow).

† When a wave is propagated from a point source, its intensity decreases as $1/R^2$, since the total energy flux is constant and is distributed over a surface which increases as R^2. The intensity, in turn, is proportional to the square of the wave amplitude.

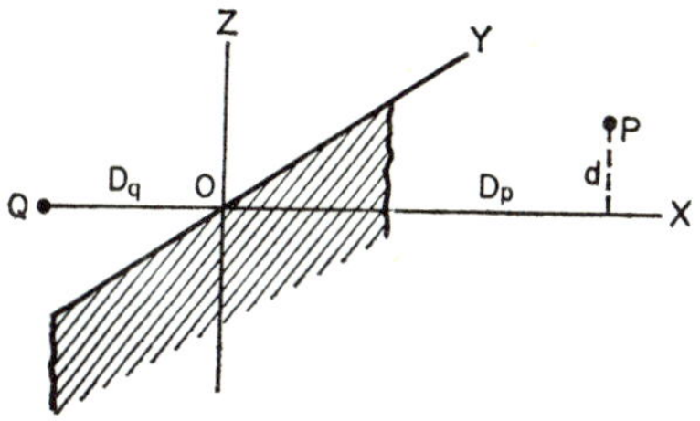

FIG. 31

We now determine the distribution of light intensity on the screen near the edge of the geometrical shadow, i.e. for values of d small compared with D_p and D_q.

As the surface of integration in (76.1) we choose the upper yz half-plane. As shown above, only the region near the origin, where y and z are small compared with D_q and D_p, makes an important contribution to the integral. For this integration it is sufficient to retain only the rapidly varying exponentials in the integrand, regarding the factor $1/R$ as constant.

The field u of the wave leaving Q is proportional to exp (ikR_q) at a distance R_q. For points on the surface of integration,

$$R_q = \sqrt{(y^2+z^2+D_q^2)} \approx D_q+\frac{1}{2D_q}(y^2+z^2),$$

and the field is

$$u \sim \exp(ikR_q) \sim \exp\left(ik\frac{y^2+z^2}{2D_q}\right).$$

The distance of the point of observation is

$$R = \sqrt{[y^2+(z-d)^2+D_p^2]} \approx D_p+\frac{1}{2D_p}[y^2+(z-d)^2].$$

We substitute this in (76.1). We then find

$$u_P \sim \int_0^\infty \exp\left\{ik\left(\frac{1}{2D_q}z^2+\frac{1}{2D_p}(z-d)^2\right)\right\}dz. \qquad (76.2)$$

Here the constant factors independent of d, including the integral with respect to y, are omitted. The integration can be extended to infinity because the integral in (76.2) is rapidly convergent, even though the form of the integrand contains the assumption that $d \ll D_q, D_p$.

The expression (76.2) can also be written in the form

$$u_P \sim \exp\left\{ik\frac{d^2}{2(D_p+D_q)}\right\}\int_0^\infty \exp\left\{ik\frac{\frac{1}{2}\left[\left(\frac{1}{D_p}+\frac{1}{D_q}\right)z-\frac{d}{D_p}\right]^2}{\frac{1}{D_p}+\frac{1}{D_q}}\right\}\mathrm{d}z.$$

The light intensity is determined by the square modulus $|u_P|^2$, which does not contain the phase factor preceding the integral. An obvious substitution reduces the integral itself to

$$u_P \sim \int_{-w}^{\infty} \mathrm{e}^{i\eta^2}\,\mathrm{d}\eta, \qquad (76.3)$$

where

$$w = d\sqrt{\frac{kD_q}{2D_p(D_q+D_p)}}. \qquad (76.4)$$

Thus, the intensity I at the point P is

$$I = \frac{I_0}{\pi}\left|\int_{-w}^{\infty} \mathrm{e}^{i\eta^2}\,\mathrm{d}\eta\right|^2; \qquad (76.5)$$

it will be shown below that I_0 is equal to the intensity in the illuminated region at points far from the edge of the shadow.

The region of geometrical shadow corresponds to negative w. Let us find the form of the function $I(w)$ for large values of $|w|$, within the shadow. Integrating by parts, we have

$$\int_{|w|}^{\infty} \mathrm{e}^{i\eta^2}\,\mathrm{d}\eta = -\frac{1}{2i|w|}\,\mathrm{e}^{iw^2}+\frac{1}{2i}\int_{|w|}^{\infty} \mathrm{e}^{i\eta^2}\,\frac{\mathrm{d}\eta}{\eta^2}.$$

Integrating by parts once more on the right side of the equation, we obtain an expression proportional to $1/|w|^3$. Thus, retaining only the term which decreases least rapidly with increasing $|w|$,

$$\int_{|w|}^{\infty} e^{i\eta^2}\, d\eta \approx -\frac{1}{2i|w|} e^{iw^2}. \tag{76.6}$$

Hence, for the intensity $I(w)$, (76.5), we obtain the following asymptotic formula, valid for large negative values of w:

$$I = I_0/4\pi w^2. \tag{76.7}$$

We now consider positive values of w, that is, the illuminated region. We write

$$\int_{-w}^{\infty} e^{i\eta^2}\, d\eta = \int_{-\infty}^{+\infty} e^{i\eta^2}\, d\eta - \int_{-\infty}^{-w} e^{i\eta^2}\, d\eta = (1+i)\sqrt{\frac{\pi}{2}} - \int_{w}^{\infty} e^{i\eta^2}\, d\eta.$$

For sufficiently large w, we can use the approximation (76.6) for the integral standing on the right-hand side of the equation, and we then easily obtain

$$I = I_0\left(1+\frac{\sin(w^2 - \frac{1}{4}\pi)}{w\sqrt{\pi}}\right). \tag{76.8}$$

Thus in the illuminated region, far from the edge of the shadow, the intensity has an infinite sequence of maxima and minima, so that the ratio I/I_0 oscillates on both sides of unity. With increasing w, the amplitude of these oscillations decreases, and the positions of the maxima and minima steadily approach one another.

For small w, the function $I(w)$ has qualitatively this same character (Fig. 32). In the region of the geometric shadow, the intensity decreases monotonically as we move away from the boundary of the shadow. (On the boundary itself, $I/I_0 = \frac{1}{4}$.) For positive w, the intensity has alternating maxima and minima. At the first (largest) maximum, $I/I_0 = 1.37$.

§77. Fraunhofer diffraction

Of special interest for physical applications are those diffraction phenomena which occur when a plane parallel bundle of rays is incident on a screen. As a result of the diffraction, the beam ceases to be parallel, and there is light propagation along directions other than the initial one. Let us consider the problem of determining the distribution over direction of the intensity of the diffracted light at large

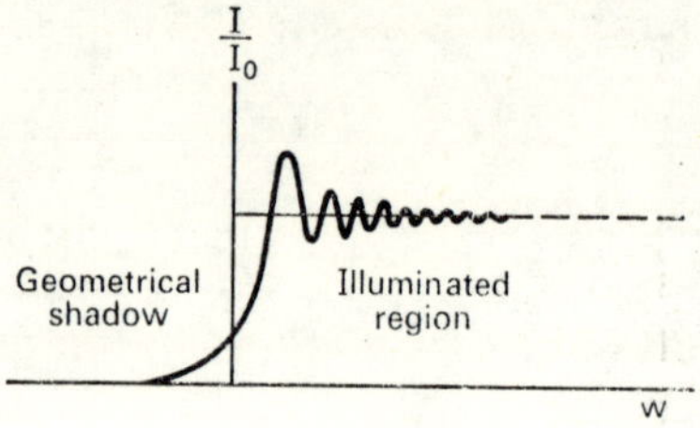

FIG. 32

distances beyond the screen (this formulation of the problem corresponds to *Fraunhofer diffraction*). Here we shall again restrict ourselves to the case of small deviations from geometrical optics, i.e. we shall assume that the angles of deviation of the rays from the initial direction (the diffraction angles) are small.†

Let us consider, for example, the Fraunhofer diffraction of a plane wave normally incident on an infinite slit (of width $2a$) with parallel sides cut in an opaque screen. We choose the plane of the slit as the yz plane, with the z axis along the slit (Fig. 33 shows a section of the screen). By symmetry, the light is deflected only in the xy plane.

Let us denote by u_0 the field which would exist beyond the screen if

† This problem can be solved by starting from the general formula (76.1) and passing to the limit where the light source and the point of observation are at infinite distances from the screen. A characteristic feature of the case we are considering is that, in the integral, the whole surface S over which the integral is taken is important (in contrast to the case of Fresnel diffraction, where only a small part of the surface is important). However, it is simpler to treat this problem anew, without recourse to the general formula (76.1), and we shall now do so.

geometrical optics were rigorously valid. It is an infinite strip of width $2a$ cut from the plane wave. Actually, however, a wave with a limited cross-sectional area cannot be strictly plane (§75). In its spatial Fourier expansion there appear components with wave vectors having different directions, and this is precisely the origin of the diffraction.

In any plane parallel to that of the slit, u_0 is non-zero and constant for $-a < y < a$. We expand this field as a Fourier integral with respect to y:

$$u_q = u_0 \int_{-a}^{a} \mathrm{e}^{-iqy}\, \mathrm{d}y = \frac{2u_0}{q} \sin qa. \tag{77.1}$$

For small deviations from geometrical optics (small q) the components in the expansion of the field u_0 can be assumed to be identical with the components of the actual field of the diffracted light, so that formula (77.1) solves our problem.

The wave vector $\mathbf{k}$ of the incident light is along the x axis. The wave vector corresponding to the component u_q of the diffracted wave is $\mathbf{k}' = \mathbf{k}+\mathbf{q}$, and lies in the xy plane at an angle θ to the x axis which is small and is such that $q \approx k\theta = \omega\theta/c$. The intensity of light diffracted in the range $\mathrm{d}q$ is proportional to $|u_q|^2\, \mathrm{d}q$. Taking u_q from (77.1), we obtain for the angular distribution of the diffracted light

$$\mathrm{d}I = \frac{I_0}{\pi ak} \frac{\sin^2 ka\theta}{\theta^2}\, \mathrm{d}\theta. \tag{77.2}$$

This is normalised so that I_0 is the total intensity, which is the same as that of the incident beam.†

$\mathrm{d}I/\mathrm{d}\theta$ as a function of the diffraction angle has the form shown in Fig. 34. As θ increases toward either side from $\theta = 0$, the intensity goes through a series of maxima with rapidly decreasing height. The

† This is easily shown by using the well-known result

$$\int_{-\infty}^{\infty} \frac{\sin^2 x}{x^2}\, \mathrm{d}x = \pi.$$

successive maxima are separated by minima at the points $\theta = n\pi/ka$ (where n is an integer); at the minima, the intensity falls to zero.

Let us consider the Fraunhofer diffraction from two screens which are "complementary", one having holes where the other is opaque: for example, an infinite slit and an infinite strip. We denote by $u_0^{(1)}$ and $u_0^{(2)}$ the fields which would exist beyond the screens according to

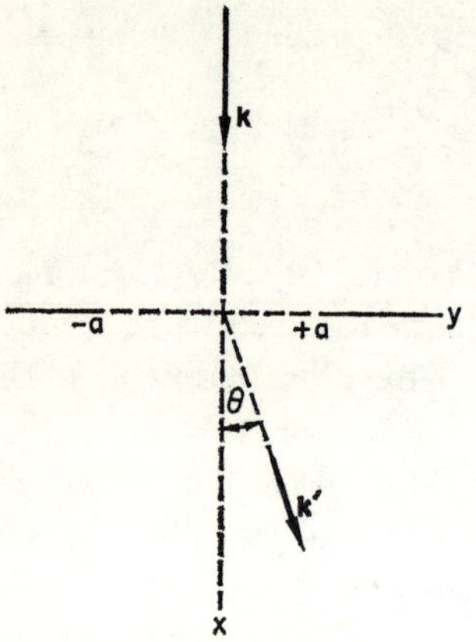

FIG. 33

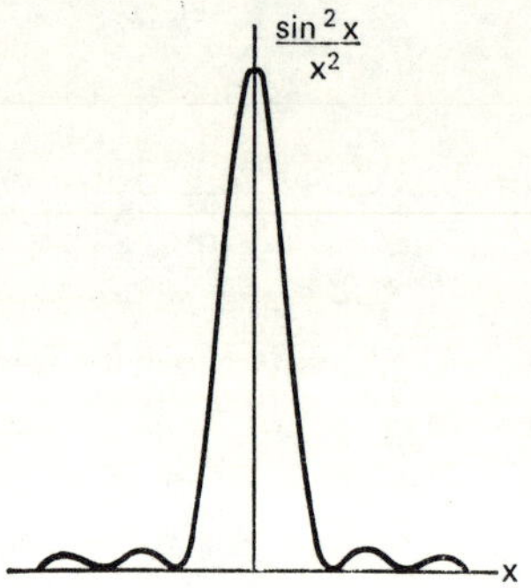

FIG. 34

geometrical optics. Since the two sets of holes in the screens together form an entire plane, the sum $u_0^{(1)}+u_0^{(2)}$ is just the total incident plane wave; this has a quite definite direction of propagation, and so its Fourier component $u_{\mathbf{q}}^{(1)}+u_{\mathbf{q}}^{(2)} = 0$ for all $\mathbf{q} \neq 0$. Hence the intensities are such that $|u_{\mathbf{q}}^{(1)}|^2 = |u_{\mathbf{q}}^{(2)}|^2$, and therefore complementary screens give the same distribution of intensity of the diffracted light (this is called *Babinet's principle*).

§78. Characteristic vibrations of the field

We consider a free electromagnetic field (no charges) in some finite volume of space. To simplify further calculations we assume that this volume has the form of a rectangular parallelepiped with sides A, B, C, respectively. Then we can expand all quantities characterising the field in this parallelepiped in a triple Fourier series (for the three coordinates). This expansion can be written (e.g. for the vector potential) in the form

$$\mathbf{A} = \sum_{\mathbf{k}} (\mathbf{a}_{\mathbf{k}} e^{i\mathbf{k}\cdot\mathbf{r}} + \mathbf{a}_{\mathbf{k}}^{*} e^{-i\mathbf{k}\cdot\mathbf{r}}), \tag{78.1}$$

explicitly indicating that **A** is real. The summation extends here over all possible values of the vector **k** whose components run through the values

$$k_x = 2\pi n_x/A, \quad k_y = 2\pi n_y/B, \quad k_z = 2\pi n_z/C, \tag{78.2}$$

where n_x, n_y, n_z are positive and negative integers. From the equation div $\mathbf{A} = 0$ it follows that, for each **k**,

$$\mathbf{k} \cdot \mathbf{a}_{\mathbf{k}} = 0, \tag{78.3}$$

i.e., the complex vectors $\mathbf{a}_{\mathbf{k}}$ are perpendicular to the corresponding wave vectors **k**. The vectors $\mathbf{a}_{\mathbf{k}}$ are, of course, functions of the time; they satisfy the equation

$$\ddot{\mathbf{a}}_{\mathbf{k}} + c^2 k^2 \mathbf{a}_{\mathbf{k}} = 0. \tag{78.4}$$

If the dimensions A, B, C of the volume are sufficiently large, then neighbouring values of k_x, k_y, k_z (for which n_x, n_y, n_z differ by unity) are very close to one another. In this case we may speak of the number of possible values of k_x, k_y, k_z in the small intervals Δk_x, Δk_y, Δk_z.

Since to neighbouring values of, say, k_x, there correspond values of n_x differing by unity, the number Δn_x of possible values of k_x in the interval Δk_x is equal simply to the number of values of n_x in the corresponding interval. Thus we obtain

$$\Delta n_x = A\,\Delta k_x/2\pi, \quad \Delta n_y = B\,\Delta k_y/2\pi, \quad \Delta n_z = C\,\Delta k_z/2\pi.$$

The total number Δn of possible values of the vector **k** with components in the intervals Δk_x, Δk_y, Δk_z is equal to the product $\Delta n_x\,\Delta n_y\,\Delta n_z$, that is,

$$\Delta n = \frac{V}{(2\pi)^3}\,\Delta k_x\,\Delta k_y\,\Delta k_z, \tag{78.5}$$

where $V = ABC$ is the volume of the field.

It is easy to determine from this the number of possible values of the wave vector having absolute values in the interval Δk, and directed into the element of solid angle Δo. To get this we need only transform to spherical polar coordinates in the "k space" and write in place of $\Delta k_x \Delta k_y \Delta k_z$ the element of volume in these coordinates. Thus

$$\Delta n = \frac{V}{(2\pi)^3} k^2 \Delta k \Delta o. \tag{78.6}$$

Finally, the number of possible values of the wave vector with absolute value k in the interval Δk and pointing in all directions is (we write 4π in place of Δo)

$$\Delta n = \frac{V}{2\pi^2} k^2 \Delta k. \tag{78.7}$$

The vectors $\mathbf{a}_\mathbf{k}$ as functions of the time behave like simply periodic functions with periods $\omega_k = ck$ (see (78.4)). We present the expansion of the field in such a form that it appears as an expansion in propagating plane waves. To do this we assume that each of the $\mathbf{a}_\mathbf{k}$ depends on the time through the factor $e^{-i\omega_k t}$:

$$\mathbf{a}_\mathbf{k} \sim e^{-i\omega_k t}, \qquad \omega_k = ck. \tag{78.8}$$

Then each individual term in the sum (78.1) is a function only of the difference $\mathbf{k}\cdot\mathbf{r}-\omega_k t$, which corresponds to a wave propagating in the direction of the vector $\mathbf{k}$.

We calculate the total energy

$$\mathcal{E} = \frac{1}{8\pi}\int (E^2+H^2)\,dV$$

of the field in the volume V, expressing it in terms of the quantities $\mathbf{a}_\mathbf{k}$. For the electric field we have

$$\mathbf{E} = -\frac{1}{c}\dot{\mathbf{A}} = -\frac{1}{c}\sum_\mathbf{k}(\dot{\mathbf{a}}_\mathbf{k} e^{i\mathbf{k}\cdot\mathbf{r}} + \dot{\mathbf{a}}^*_\mathbf{k} e^{-i\mathbf{k}\cdot\mathbf{r}}),$$

or, keeping in mind (78.8),

$$\mathbf{E} = i\sum_\mathbf{k} k(\mathbf{a}_\mathbf{k} e^{i\mathbf{k}\cdot\mathbf{r}} - \mathbf{a}^*_\mathbf{k} e^{-i\mathbf{k}\cdot\mathbf{r}}). \tag{78.9}$$

For the magnetic field $\mathbf{H} = \text{curl}\,\mathbf{A}$, we obtain

$$\mathbf{H} = i\sum_{\mathbf{k}}(\mathbf{k}\times\mathbf{a}_{\mathbf{k}}e^{i\mathbf{k}\cdot\mathbf{r}} - \mathbf{k}\times\mathbf{a}_{\mathbf{k}}^{*}e^{-i\mathbf{k}\cdot\mathbf{r}}). \tag{78.10}$$

When calculating the squares of these sums, we must keep in mind that all products of terms with wave vectors $\mathbf{k} \neq \mathbf{k}'$ give zero on integration over the whole volume, since such terms contain factors of the form $e^{\pm i\mathbf{q}\cdot\mathbf{r}}$, $\mathbf{q} = \mathbf{k}\pm\mathbf{k}'$, and the integral, e.g.

$$\int_0^A e^{i(2\pi/A)n_x x}\,dx,$$

with integer n_x different from zero, gives zero. In those terms from which the exponentials drop out, integration over dV gives just the volume V.

As a result, we obtain

$$\mathscr{E} = \frac{V}{4\pi}\sum_{\mathbf{k}}\{k^2\mathbf{a}_{\mathbf{k}}\cdot\mathbf{a}_{\mathbf{k}}^{*} + (\mathbf{k}\times\mathbf{a}_{\mathbf{k}})\cdot(\mathbf{k}\times\mathbf{a}_{\mathbf{k}}^{*})\}.$$

Since $\mathbf{a}_{\mathbf{k}}\cdot\mathbf{k} = 0$, we have

$$(\mathbf{k}\times\mathbf{a}_{\mathbf{k}})\cdot(\mathbf{k}\times\mathbf{a}_{\mathbf{k}}^{*}) = k^2\mathbf{a}_{\mathbf{k}}\cdot\mathbf{a}_{\mathbf{k}}^{*},$$

and we obtain finally

$$\mathscr{E} = \sum_{\mathbf{k}}\mathscr{E}_{\mathbf{k}}, \quad \mathscr{E}_{\mathbf{k}} = \frac{k^2V}{2\pi}\mathbf{a}_{\mathbf{k}}\cdot\mathbf{a}_{\mathbf{k}}^{*}. \tag{78.11}$$

Thus the total energy of the field is expressed as a sum of the energies $\mathscr{E}_{\mathbf{k}}$ associated with each of the plane waves individually.

In a completely analogous fashion, we can calculate the total momentum of the field,

$$\frac{1}{c^2}\int\mathbf{S}\,dV = \frac{1}{4\pi c}\int\mathbf{E}\times\mathbf{H}\,dV,$$

for which we obtain

$$\sum_{\mathbf{k}}\frac{\mathbf{k}}{k}\frac{\mathscr{E}_{\mathbf{k}}}{c}. \tag{78.12}$$

This result could have been anticipated in view of the relation between the energy and momentum of a plane wave (see §69).

The expansion (78.1) succeeds in expressing the field in terms of a series of discrete variables (the vectors $\mathbf{a}_\mathbf{k}$), in place of the description in terms of a continuous series of variables, which is essentially what is done when we give the potential $\mathbf{A}(x, y, z, t)$ at all points of space. We now make a transformation of the variables $\mathbf{a}_\mathbf{k}$, which has the result that the equations of the field take on a form similar to the canonical equations (Hamilton's equations) of mechanics.

We introduce the real "canonical variables" $\mathbf{Q}_\mathbf{k}$ and $\mathbf{P}_\mathbf{k}$ according to the relations

$$\left.\begin{aligned} \mathbf{Q}_\mathbf{k} &= \sqrt{\frac{V}{4\pi c^2}}(\mathbf{a}_\mathbf{k}+\mathbf{a}_\mathbf{k}^*), \\ \mathbf{P}_\mathbf{k} &= -i\omega_k\sqrt{\frac{V}{4\pi c^2}}(\mathbf{a}_\mathbf{k}-\mathbf{a}_\mathbf{k}^*) = \dot{\mathbf{Q}}_\mathbf{k}. \end{aligned}\right\} \tag{78.13}$$

The Hamiltonian of the field is obtained by substituting these expressions in the energy (78.11):

$$\mathscr{H} = \sum_\mathbf{k} \mathscr{H}_\mathbf{k} = \sum_\mathbf{k} \tfrac{1}{2}(\mathbf{P}_\mathbf{k}^2+\omega_k^2\mathbf{Q}_\mathbf{k}^2). \tag{78.14}$$

Then Hamilton's equations $\partial\mathscr{H}/\partial\mathbf{P}_\mathbf{k} = \dot{\mathbf{Q}}_\mathbf{k}$ coincide with $\mathbf{P}_\mathbf{k} = \dot{\mathbf{Q}}_\mathbf{k}$, which is thus a consequence of the equations of motion. (This was achieved by an appropriate choice of the coefficient in (78.13).) The equations of motion $\partial\mathscr{H}/\partial\mathbf{Q}_\mathbf{k} = -\dot{\mathbf{P}}_\mathbf{k}$ become the equations

$$\ddot{\mathbf{Q}}_\mathbf{k}+\omega_k^2\mathbf{Q}_\mathbf{k} = 0, \tag{78.15}$$

that is, they are identical with the equations of the field.

Each of the vectors $\mathbf{Q}_\mathbf{k}$ and $\mathbf{P}_\mathbf{k}$ is perpendicular to the wave vector $\mathbf{k}$, i.e. has two independent components. The direction of these vectors determines the direction of polarisation of the corresponding travelling wave. Denoting the two components of the vector $\mathbf{Q}_\mathbf{k}$ (in the plane perpendicular to $\mathbf{k}$) by $Q_{\mathbf{k}j}$, $j = 1, 2$, we have

$$\mathbf{Q}_\mathbf{k}^2 = \sum_j Q_{\mathbf{k}j}^2,$$

and similarly for $\mathbf{P}_\mathbf{k}$. Then

$$\mathscr{H} = \sum_{\mathbf{k}j} \mathscr{H}_{\mathbf{k}j}, \quad \mathscr{H}_{\mathbf{k}j} = \tfrac{1}{2}(P_{\mathbf{k}j}^2+\omega_k^2 Q_{\mathbf{k}j}^2). \tag{78.16}$$

We see that the Hamiltonian splits into a sum of independent terms $\mathscr{H}_{\mathbf{k}j}$, each of which contains only one pair of the quantities $Q_{\mathbf{k}j}$, $P_{\mathbf{k}j}$. Each such term corresponds to a travelling wave with a definite wave vector and polarisation. The quantity $\mathscr{H}_{\mathbf{k}j}$ has the form of the Hamiltonian of a one-dimensional "oscillator", executing a simple harmonic vibration. For this reason, one sometimes refers to this result as the expansion of the field in terms of *oscillators*.

CHAPTER 14

RADIATION OF ELECTROMAGNETIC WAVES

§79. The retarded potentials

Let us derive the equations determining the potentials of the field produced by moving charges. To do this we repeat the derivation given in §68, but now do not assume that the charge and current densities are zero.

Substituting the definitions

$$\mathbf{E} = -\frac{1}{c}\frac{\partial \mathbf{A}}{\partial t} - \nabla\phi, \quad \mathbf{H} = \operatorname{curl} \mathbf{A} \tag{79.1}$$

in the equation

$$\operatorname{curl} \mathbf{H} = \frac{4\pi}{c}\mathbf{j} + \frac{1}{c}\frac{\partial \mathbf{E}}{\partial t},$$

we get

$$\operatorname{curl}\operatorname{curl} \mathbf{A} = -\triangle \mathbf{A} + \operatorname{grad}\operatorname{div} \mathbf{A} = \frac{4\pi}{c}\mathbf{j} - \frac{1}{c^2}\frac{\partial^2 \mathbf{A}}{\partial t^2} - \operatorname{grad}\frac{\partial \phi}{\partial t} \tag{79.2}$$

(where we have interchanged the order of the operations grad and $\partial/\partial t$ in the last term).

As the supplementary condition on the potentials we now choose

$$\operatorname{div} \mathbf{A} + \frac{1}{c}\frac{\partial \phi}{\partial t} = 0. \tag{79.3}$$

This condition is called the Lorentz condition, and the potentials satisfying it are said to be in the *Lorentz gauge.*† Then the last terms on both sides of (79.2) cancel and we arrive at the equation

$$\triangle \mathbf{A} - \frac{1}{c^2}\frac{\partial^2 \mathbf{A}}{\partial t^2} = -\frac{4\pi}{c}\mathbf{j}. \tag{79.4}$$

Similarly, substituting (79.1) in the equation div $\mathbf{E} = 4\pi\varrho$, we get

$$-\frac{1}{c}\frac{\partial}{\partial t}\operatorname{div}\mathbf{A} - \triangle\phi = 4\pi\varrho,$$

or, expressing div **A** from the condition (79.3),

$$\triangle\phi - \frac{1}{c^2}\frac{\partial^2\phi}{\partial t^2} = -4\pi\varrho. \tag{79.5}$$

Equations (79.4), (79.5) are the desired equations. For constan fields, these reduce to the already familiar equations (59.4) and (65.4) and for variable fields without charges, to the homogeneous wave equation.

As we know, the solution of the inhomogeneous linear equations (79.4) and (79.5) can be represented as the sum of the solution of these equations without the right-hand side, and a particular integral of these equations with the right-hand side. To find the particular solution, we divide the whole space into infinitely small regions and determine the field produced by a charge located in one of these volume elements. Because of the linearity of the field equations, the actual field will be the sum of the fields produced by all such elements.

The charge de in a given volume element is, generally speaking, a function of the time. If we choose the origin of coordinates in the volume element under consideration, then the charge density is

---

† The condition (79.3) is more general than the conditions $\phi = 0$, div $\mathbf{A} = 0$ that were used in §68; potentials that satisfy the latter conditions also satisfy (79.3). Unlike those conditions, the Lorentz condition has an invariant character: potentials that satisfy it in one frame of reference also satisfy it in every other frame of reference. This is clear from the fact that the condition (79.3) can be written in a four-dimensional invariant form:

$$\partial A^{\mu}/\partial x^{\mu} = 0.$$

$\varrho = de(t)\, \delta(\mathbf{R})$, where $\mathbf{R}$ is the distance from the origin. Thus we must solve the equation

$$\triangle\phi - \frac{1}{c^2}\frac{\partial^2\phi}{\partial t^2} = -4\pi\, de(t)\,\delta(\mathbf{R}). \tag{79.6}$$

Everywhere, except at the origin, $\delta(\mathbf{R}) = 0$, and we have the equation

$$\triangle\phi - \frac{1}{c^2}\frac{\partial^2\phi}{\partial t^2} = 0. \tag{79.7}$$

It is clear that in the case we are considering ϕ has central symmetry, i.e., ϕ is a function only of R. Therefore if we write the Laplace operator in spherical coordinates, (79.7) reduces to

$$\frac{1}{R^2}\frac{\partial}{\partial R}\left(R^2\frac{\partial\phi}{\partial R}\right) - \frac{1}{c^2}\frac{\partial^2\phi}{\partial t^2} = 0.$$

To solve this equation, we make the substitution $\phi = \chi(R, t)/R$. Then, we find for χ

$$\frac{\partial^2\chi}{\partial R^2} - \frac{1}{c^2}\frac{\partial^2\chi}{\partial t^2} = 0.$$

But this is the equation of plane waves, whose solution has the form

$$\chi = f_1[t-(R/c)] + f_2[t+(R/c)].$$

Since we only want a particular solution of the equation, it is sufficient to choose only one of the functions f_1 and f_2. Usually it turns out to be convenient to take $f_2 = 0$ (concerning this, see below). Then, everywhere except at the origin, ϕ has the form

$$\phi = \frac{\chi[t-(R/c)]}{R}. \tag{79.8}$$

So far the function χ is arbitrary; we now choose it so that we also obtain the correct value for the potential at the origin In other words, we must select χ so that at the origin equation (79.6) is satisfied. This is easily done, noting that as $R \to 0$, the potential increases to infinity, and therefore its derivatives with respect to the coordinates increase more rapidly than its time derivatives. Consequently as $R \to 0$, we can, in equation (79.6), neglect the term $(1/c^2)(\partial^2\phi/\partial t^2)$ compared with $\triangle\phi$.

Then it goes over into the familiar equation (59.10) leading to Coulomb's law. Thus, near the origin, (79.8) must go over into Coulomb's law, from which it follows that $\chi(t) = \mathrm{d}e(t)$, that is,

$$\phi = \frac{\mathrm{d}e[t-(R/c)]}{R}.$$

From this it is easy to get to the solution of equation (79.5) for an arbitrary distribution of charges $\varrho(x, y, z, t)$. To do this, it is sufficient to write $\mathrm{d}e = \varrho \, \mathrm{d}V$ ($\mathrm{d}V$ is the volume element) and integrate over the whole space. To this solution of the inhomogeneous equation (79.5) we can still add the solution ϕ_0 of the same equation without the right-hand side. Thus, the general solution has the form

$$\phi(\mathbf{r}, t) = \int \frac{1}{R} \varrho[\mathbf{r}', t-(R/c)] \, \mathrm{d}V' + \phi_0, \tag{79.9}$$

$$\mathbf{R} = \mathbf{r} - \mathbf{r}', \quad \mathrm{d}V' = \mathrm{d}x' \, \mathrm{d}y' \, \mathrm{d}z'$$

where

$$\mathbf{r} = (x, y, z), \quad \mathbf{r}' = (x', y', z');$$

R is the distance from the volume element $\mathrm{d}V'$ to the "field point" at which we determine the potential. We shall write this expression briefly as

$$\phi = \int \frac{\varrho_{t-R/c}}{R} \, \mathrm{d}V + \phi_0, \tag{79.10}$$

where the subscript means that the quantity ϱ is to be taken at the time $t-(R/c)$, and the prime on $\mathrm{d}V$ has been omitted.

Similarly we have for the vector potential

$$\mathbf{A} = \frac{1}{c} \int \frac{\mathbf{j}_{t-R/c}}{R} \, \mathrm{d}V + \mathbf{A}_0, \tag{79.11}$$

where $\mathbf{A}_0$ is the solution of equation (79.4) without the right-hand term.

The potentials (79.10) and (79.11) (without ϕ_0 and $\mathbf{A}_0$) are called the *retarded potentials*.

If the charges are at rest (i.e. density ϱ independent of the time), formula (79.10) goes over into the known formula (59.9) for the electrostatic field potential; for the case of stationary motion of the charges, formula (79.11), after averaging, goes over into formula (65.5) for the vector potential of a constant magnetic field.

The quantities $\mathbf{A}_0$ and ϕ_0 in (79.10) and (79.11) are to be determined so that the conditions of the problem are fulfilled. To do this it is clearly sufficient to impose initial conditions, that is, to fix the values of the field at the initial time. However, we do not usually have to deal with such initial conditions. Instead we are usually given conditions at large distances from the system of charges throughout all time. Thus, we may be told that radiation is incident on the system from outside. Corresponding to this, the field which is developed as a result of the interaction of this radiation with the system can differ from the external field only by the radiation originating from the system. This radiation emitted by the system must, at large distances, have the form of waves spreading out from the system, that is, in the direction of increasing R. But precisely this condition is satisfied by the retarded potentials. Thus these solutions represent the field produced by the system, while ϕ_0 and $\mathbf{A}_0$ must be put equal to the external field acting on the system.

§80. The Lienard–Wiechert potentials

Let us determine the potentials for the field produced by a point charge executing an assigned motion along a path $\mathbf{r} = \mathbf{r}_0(t)$.

According to the formulae for the retarded potentials, the field at the point of observation $P(x, y, z)$ at time t is determined by the state of motion of the charge at the earlier time t', for which the time of propagation of the light signal from the point $\mathbf{r}_0(t')$, where the charge was located, to the field point P just coincides with the difference $t-t'$. Let $\mathbf{R}(t) = \mathbf{r}-\mathbf{r}_0(t)$ be the radius vector from the charge e to the point P; like $\mathbf{r}_0(t)$ it is a given function of the time. Then the time t' is determined by the equation

$$t'+\frac{R(t')}{c} = t. \tag{80.1}$$

In the frame of reference in which the particle is at rest at time t', the potential at the point of observation at time t is just the Coulomb potential,

$$\phi = \frac{e}{R(t')}, \quad \mathbf{A} = 0. \tag{80.2}$$

The expressions for the potentials in an arbitrary reference frame can be found directly by finding a four-vector which for $\mathbf{v} = 0$ coincides with the expressions just given for ϕ and $\mathbf{A}$. Noting that, according to (80.1), ϕ in (80.2) can also be written in the form

$$\phi = \frac{e}{c(t-t')},$$

we find that the required four-vector is

$$A^\mu = e\frac{u^\mu}{(R_\nu u^\nu)}, \tag{80.3}$$

where u^μ is the four-velocity of the charge, $R^\nu = [c(t-t'),\ \mathbf{r}-\mathbf{r}']$, where x', y', z', t' are related by the equation (80.1). This equation has an invariant character, since it can be written in invariant form:

$$R_\nu R^\nu = 0. \tag{80.4}$$

Now once more transforming to three-dimensional notation, we obtain, for the potentials of the field produced by an arbitrarily moving point charge, the following expressions:

$$\phi = \frac{e}{R-\mathbf{v}\cdot\mathbf{R}/c}, \quad \mathbf{A} = \frac{e\mathbf{v}}{c(R-\mathbf{v}\cdot\mathbf{R}/c)}, \tag{80.5}$$

where $\mathbf{R}$ is the radius vector, taken from the point where the charge is located to the point of observation P, and all the quantities on the right-hand sides of the equations must be evaluated at the time t', determined from (80.1). The potentials of the field, in the form (80.5), are called the *Lienard–Wiechert potentials.*

To calculate the intensities of the electric and magnetic fields from the formulae

$$\mathbf{E} = -\frac{1}{c}\frac{\partial \mathbf{A}}{\partial t} - \operatorname{grad}\phi, \quad \mathbf{H} = \operatorname{curl}\mathbf{A},$$

we must differentiate ϕ and **A** with respect to the coordinates x, y, z of the point, and the time t of observation. But the formulae (80.5) express the potentials as functions of t', and only through the relation (80.1) as implicit functions of x, y, z, t. Therefore to calculate the required derivatives we must first calculate the derivatives of t'. Differentiating the relation $R(t') = c(t-t')$ with respect to t and $\mathbf{r}$, we get

$$\left.\begin{aligned} \frac{\partial R}{\partial t} &= \frac{\partial R}{\partial t'}\frac{\partial t'}{\partial t} = c\left(1-\frac{\partial t'}{\partial t}\right), \\ \operatorname{grad} R \equiv \frac{\partial R}{\partial \mathbf{r}} &= \frac{\partial R}{\partial t'}\operatorname{grad} t' + \frac{\partial R}{\partial \mathbf{r}} = -c \operatorname{grad} t'. \end{aligned}\right\} \quad (80.6)$$

We find the derivative $\partial R/\partial t'$ by differentiating the identity $R^2 = \mathbf{R}^2$ and substituting $\partial \mathbf{R}/\partial t' = -\mathbf{v}(t')$ (where the minus sign occurs because $\mathbf{R}$ is the radius vector from the charge e to the point P, while the velocity is the time derivative of the coordinates of the charge); then

$$\partial R/\partial t' = -\mathbf{R}\cdot\mathbf{v}/R.$$

The other derivative is

$$\partial R/\partial \mathbf{r} = \mathbf{R}/R.$$

Substituting these values in (80.6), we find

$$\frac{\partial t'}{\partial t} = \frac{1}{1-\mathbf{v}\cdot\mathbf{R}/cR}, \quad \operatorname{grad} t' = -\frac{\mathbf{R}}{c(R-\mathbf{R}\cdot\mathbf{v}/c)}. \quad (80.7)$$

With the aid of these formulae, there is no difficulty in carrying out the calculation of the fields **E** and **H**. Omitting the intermediate calculations, we give the final results:

$$\mathbf{E} = e\frac{[1-(v^2/c^2)]}{(R-\mathbf{R}\cdot\mathbf{v}/c)^3}\left(\mathbf{R}-\frac{\mathbf{v}}{c}R\right) + \frac{e}{c^2(R-\mathbf{R}\cdot\mathbf{v}/c)^3}\mathbf{R}\times\left\{\left(\mathbf{R}-\frac{\mathbf{v}}{c}R\right)\times\dot{\mathbf{v}}\right\}, \quad (80.8)$$

$$\mathbf{H} = \frac{1}{R}\mathbf{R}\times\mathbf{E}. \quad (80.9)$$

Here, $\dot{\mathbf{v}} = \partial\mathbf{v}/\partial t'$; all quantities on the right-hand sides of the equations refer to the time t'. It is interesting to note that the magnetic field turns out to be everywhere perpendicular to the electric field.

The electric field (80.8) consists of two parts of different type. The first term depends only on the velocity of the particle (and not on its acceleration) and varies at large distances like $1/R^2$. The second term depends on the acceleration, and for large R it varies like $1/R$. Later we shall see that this latter term is related to the electromagnetic waves radiated by the particle.

As for the first term, since it is independent of the acceleration it must correspond to the field produced by a uniformly moving charge. In fact, one can show (we omit the proof) that the field given by this term is identical with the field (61.5).

§81. The field of a system of charges at large distances

Let us consider the field produced by a system of moving charges at distances large compared with the dimensions of the system.

We choose the origin of coordinates O anywhere in the interior of the system of charges. The position vector from O to the point P, where we determine the field, we denote by $\mathbf{R}_0$, and the unit vector in this direction by $\mathbf{n}$. Let the position vector of the charge element $de = \varrho\, dV$ be $\mathbf{r}$, and the radius vector from de to the point P be $\mathbf{R}$. Obviously $\mathbf{R} = \mathbf{R}_0 - \mathbf{r}$.

At large distances from the system of charges, $R_0 \gg r$, and we have approximately

$$R = |\mathbf{R}_0 - \mathbf{r}| \approx R_0 - \mathbf{r}\cdot\mathbf{n}.$$

We substitute this in formulae (79.10), (79.11) for the retarded potentials. In the denominator of the integrands we can neglect $\mathbf{r}\cdot\mathbf{n}$ compared with R_0. In $t - R/c$, however, this is generally not permissible; whether it is possible to neglect these terms is determined not by the relative values of R_0/c and $\mathbf{r}\cdot\mathbf{n}/c$, but by how much the quantities ϱ and $\mathbf{j}$ change during the time $\mathbf{r}\cdot\mathbf{n}/c$. Since R_0 is constant in the integration and can be taken out from under the integral sign, we get for the potentials of the field at large distances from the system of charges the expressions

$$\phi = \frac{1}{R_0} \int \varrho_{t-R_0/c+\mathbf{r}\cdot\mathbf{n}/c}\,\mathrm{d}V, \tag{81.1}$$

$$\mathbf{A} = \frac{1}{cR_0} \int \mathbf{j}_{t-R_0/c+\mathbf{r}\cdot\mathbf{n}/c}\,\mathrm{d}V. \tag{81.2}$$

At sufficiently large distances from the system of charges, the field over small regions of space can be considered to be a plane wave. For this it is necessary that the distance be large compared not only with the dimensions of the system, but also with the wavelength of the electromagnetic waves radiated by the system. We refer to this region of space as the *wave zone* of the radiation.

In a plane wave, the fields **E** and **H** are related to each other by (69.4), $\mathbf{E} = \mathbf{H}\times\mathbf{n}$. Since $\mathbf{H} = \operatorname{curl} \mathbf{A}$, it is sufficient for a complete determination of the field in the wave zone to calculate only the vector potential. In a plane wave we have $\mathbf{H} = (1/c)\dot{\mathbf{A}}\times\mathbf{n}$ [see (69.3)], where the dot indicates differentiation with respect to time. Thus, knowing **A**, we find **H** and **E** from the formulae†

$$\mathbf{H} = \frac{1}{c}\dot{\mathbf{A}}\times\mathbf{n}, \quad \mathbf{E} = \frac{1}{c}(\dot{\mathbf{A}}\times\mathbf{n})\times\mathbf{n}. \tag{81.3}$$

We note that the field at large distances is inversely proportional to the first power of the distance R_0 from the radiating system. We also note that the time t enters into the expressions (81.1) – (81.3) always in the combination $t-R_0/c$.

The radiated electromagnetic waves carry off energy. The energy flux is given by the Poynting vector, which, for a plane wave, is

$$\mathbf{S} = c\frac{H^2}{4\pi}\mathbf{n}.$$

The intensity $\mathrm{d}I$ of radiation into the element of solid angle $\mathrm{d}o$ is

† The formula $\mathbf{E} = -(1/c)\dot{\mathbf{A}}$ [see (69.3)] is here not applicable to the potentials ϕ, **A**, since they do not satisfy the same auxiliary conditions as were imposed on them in §69.

defined as the amount of energy passing in unit time through the element $df = R_0^2\,do$ of the spherical surface with centre at the origin and radius R_0. This quantity is clearly equal to the energy flux density S multiplied by df, i.e.

$$dI = c\frac{H^2}{4\pi}R_0^2\,do. \tag{81.4}$$

Since the field H is inversely proportional to R_0, we see that the amount of energy radiated by the system in unit time into the element of solid angle do is the same for all distances (if the values of $t-R_0/c$ are the same for them). This is, of course, as it should be, since the energy radiated from the system spreads out with velocity c into the surrounding space, not accumulating or disappearing anywhere.

§82. Dipole radiation

The time $\mathbf{r}\cdot\mathbf{n}/c$ in the integrands of the expressions (81.1) and (81 2) for the retarded potentials can be neglected in cases where the distribution of charge changes little during this time. It is easy to find the conditions for satisfying this requirement. Let T denote the order of magnitude of the time during which the distribution of the charges in the system changes significantly. The radiation of the system will obviously contain periods of order T (i.e. frequencies of order $1/T$). We further denote by a the order of magnitude of the dimensions of the system. Then the time $\mathbf{r}\cdot\mathbf{n}/c \sim a/c$. In order that the distribution of the charges in the system shall not undergo a significant change during this time, it is necessary that $a/c \ll T$. But cT is just the wavelength λ of the radiation. Thus the condition $a \ll cT$ can be written in the form

$$a \ll \lambda, \tag{82.1}$$

that is, the dimensions of the system must be small compared with the radiated wavelength.

This condition can be written in still another form by noting that

$T \sim a/v$, so that $\lambda \sim ca/v$, if v is of the order of magnitude of the velocities of the charges. From $a \ll \lambda$, we then find

$$v \ll c, \tag{82.2}$$

that is, the velocities of the charges must be small compared with the velocity of light.

We shall assume that this condition is fulfilled, and consider the radiation at distances from the radiating system large compared with the wavelength (and consequently, in any case, large compared with the dimensions of the system). As was pointed out in §81, at such distances the field can be considered as a plane wave, and therefore in determining the field it is sufficient to calculate only the vector potential.

The vector potential (81.2) of the field now has the form

$$\mathbf{A} = \frac{1}{cR_0}\int \mathbf{j}_{t'}\,\mathrm{d}V, \tag{82.3}$$

where the time $t' = t - R_0/c$ now no longer depends on the variable of integration. Substituting $\mathbf{j} = \varrho\mathbf{v}$, we rewrite (82.3) in the form

$$\mathbf{A} = \frac{1}{cR_0}\left(\sum e\mathbf{v}\right)$$

(the summation goes over all the charges of the system; for brevity, we omit the index t'—all quantities on the right-hand side of the equation refer to time t'). But

$$\sum e\mathbf{v} = \frac{\mathrm{d}}{\mathrm{d}t}\sum e\mathbf{r} = \dot{\mathbf{d}},$$

where $\mathbf{d}$ is the dipole moment of the system. Thus,

$$\mathbf{A} = \frac{1}{cR_0}\dot{\mathbf{d}}. \tag{82.4}$$

With the aid of formulae (81.3) we find that the magnetic field is equal to

$$\mathbf{H} = \frac{1}{c^2R_0}\ddot{\mathbf{d}}\times\mathbf{n}, \tag{82.5}$$

and the electric field to

$$\mathbf{E} = \frac{1}{c^2 R_0} (\ddot{\mathbf{d}} \times \mathbf{n}) \times \mathbf{n}. \tag{82.6}$$

We note that, in the approximation considered here, the radiation is determined by the second derivative of the dipole moment of the system. Radiation of this kind is called *dipole radiation.*

Since $\mathbf{d} = \Sigma e\mathbf{r}$, $\ddot{\mathbf{d}} = \Sigma e\dot{\mathbf{v}}$. Thus the charges can radiate only if they move with acceleration. Charges in uniform motion do not radiate. This also follows directly from the principle of relativity, since a charge in uniform motion can be considered in the inertial frame in which it is at rest, and a charge at rest does not radiate.

Substituting (82.5) in (81.4), we get the intensity of the dipole radiation:

$$\mathrm{d}I = \frac{1}{4\pi c^3} (\ddot{\mathbf{d}} \times \mathbf{n})^2 \, \mathrm{d}o = \frac{\ddot{\mathbf{d}}^2}{4\pi c^3} \sin^2 \theta \, \mathrm{d}o, \tag{82.7}$$

where θ is the angle between $\ddot{\mathbf{d}}$ and $\mathbf{n}$. This is the amount of energy radiated by the system in unit time into the element of solid angle do. We note that the angular distribution of the radiation is given by the factor $\sin^2 \theta$.

Substituting d$o = 2\pi \sin \theta \, \mathrm{d}\theta$ and integrating over θ from 0 to π, we find for the total radiation

$$I = \frac{2}{3c^3} \ddot{\mathbf{d}}^2. \tag{82.8}$$

If we have just one charge moving in the external field, then $\mathbf{d} = e\mathbf{r}$ and $\ddot{\mathbf{d}} = e\mathbf{w}$, where $\mathbf{w}$ is the acceleration of the charge. Thus the total radiation of the moving charge is

$$I = \frac{2e^2 w^2}{3c^3}. \tag{82.9}$$

We can perform a spectral resolution of the radiation emitted by the system. It is obvious that the emission of a particular monochromatic component of the radiation is due to the corresponding component of the dipole moment of the system, $\mathbf{d}(t)$. In treating the problem we must distinguish the cases of expansion in a Fourier series or in a Fourier integral.

If the charges execute a periodic motion (with frequency ω_0), then the dipole moment (and with it the radiation field) must be expanded in Fourier series. According to the general formula (72.4) the intensity of a monochromatic component (with frequency $\omega = n\omega_0$) is obtained from the formula for the average intensity of the radiation,

$$I = \frac{2}{3c^3}\overline{\ddot{\mathbf{d}}^2}, \tag{82.10}$$

by replacing the mean square $\overline{\ddot{\mathbf{d}}^2}$ by twice the squared modulus of the corresponding Fourier component:

$$I_n = \frac{4}{3c^3}|\ddot{\mathbf{d}}_n|^2.$$

We can express the Fourier components of the vector $\ddot{\mathbf{d}}(t)$ in terms of the Fourier components of the vector $\mathbf{d}(t)$. To do this we note that each term in the expansion of $\ddot{\mathbf{d}}(t)$ as a series or as an integral must be obtained by a time differentiation of the corresponding term in the expansion of $\mathbf{d}(t)$, i.e.,

$$\ddot{\mathbf{d}}_\omega \mathrm{e}^{-i\omega t} = \frac{\mathrm{d}^2}{\mathrm{d}t^2}(\mathbf{d}_\omega \mathrm{e}^{-i\omega t}) = -\omega^2\, \mathbf{d}_\omega \mathrm{e}^{-i\omega t},$$

so that

$$\ddot{\mathbf{d}}_\omega = -\omega^2\, \mathbf{d}_\omega. \tag{82.11}$$

Therefore

$$I_n = \frac{4n^4\omega_0^4}{3c^3}|\mathbf{d}_n|^2. \tag{82.12}$$

We have to work with the Fourier integral expansion in the case of radiation accompanying the collision of charged particles (*bremsstrahlung*). Here the quantity of interest is the total amount of energy radiated during the course of the collision. Let $\mathrm{d}\mathscr{E}_\omega$ be the energy radiated in the form of waves with frequencies in the interval between ω and $\omega+\mathrm{d}\omega$. According to (72.8) we obtain it from the formula for the total energy of the radiation,

$$\Delta\mathscr{E} = \int_{-\infty}^{\infty} I\,\mathrm{d}t = \frac{2}{3c^3}\int_{-\infty}^{\infty} \ddot{\mathbf{d}}^2\,\mathrm{d}t, \tag{82.13}$$

by replacing the integral by the expression $2|\ddot{\mathbf{d}}_\omega|^2\,\mathrm{d}\omega/2\pi$:

$$\mathrm{d}\mathscr{E}_\omega = \frac{2}{3\pi c^3}|\ddot{\mathbf{d}}_\omega|^2\,\mathrm{d}\omega = \frac{2\omega^4}{3\pi c^3}|\mathbf{d}_\omega|^2\,\mathrm{d}\omega. \tag{82.14}$$

We note that a closed system consisting of particles all with the same charge-to-mass ratio cannot emit dipole radiation: for such a system the dipole moment

$$\mathbf{d} = \sum e\mathbf{r} = \sum \frac{e}{m} m\mathbf{r} = \text{constant}\times\sum m\mathbf{r},$$

where the constant is the charge-to-mass ratio common to all the particles. But $\Sigma m\mathbf{r} = \mathbf{R}\,\Sigma m$, where $\mathbf{R}$ is the position vector of the centre of mass of the system (all velocities $v \ll c$, so that non-relativistic mechanics is applicable). Thus $\ddot{\mathbf{d}}$ is proportional to the acceleration of the centre of mass, and is thus zero, since the centre of mass moves uniformly.

If dipole radiation is absent, we must go to higher terms in the expansion of the potentials in powers of a/λ in order to find the energy radiated by the system. In the next approximation (after the dipole radiation) we find radiation resulting from the oscillations of both the electric quadrupole moment of the system and its magnetic moment.

PROBLEMS†

PROBLEM 1. Find the radiation from a dipole $\mathbf{d}$, rotating in a plane with constant angular velocity Ω.

SOLUTION. Choosing the plane of the rotation as the xy plane, we have

$$d_x = d_0\cos\Omega t, \quad d_y = d_0\sin\Omega t.$$

Since these functions are monochromatic, the radiation is also monochromatic, with frequency $\omega = \Omega$. From formula (82.7) we find for the angular distribution

† In all the problems it is assumed that all particle velocities $v \ll c$.

of the radiation (averaged over the period of the rotation)

$$\overline{dI} = \frac{d_0^2 \Omega^4}{8\pi c^3}(1+\cos^2\vartheta)\,do,$$

where ϑ is the angle between the direction $\mathbf{n}$ of the radiation and the z axis. The total intensity is

$$\overline{I} = \frac{2d_0^2\Omega^4}{3c^3}.$$

Problem 2. Determine the total radiation in the head-on collision of two particles repelling one another.

Solution. Choosing the coordinate origin at the centre of mass of the particles, we find for the dipole moment of the system

$$\mathbf{d} = e_1\mathbf{r}_1 + e_2\mathbf{r}_2 = \frac{e_1 m_2 - e_2 m_1}{m_1+m_2}\mathbf{r} = \mu\left(\frac{e_1}{m_1} - \frac{e_2}{m_2}\right)\mathbf{r},$$

where the subscripts 1 and 2 refer to the two particles, $\mathbf{r} = \mathbf{r}_1 - \mathbf{r}_2$ is the radius vector joining them, and $\mu = m_1 m_2/(m_1+m_2)$ is their reduced mass. The equation of the relative motion of the particles is

$$\dot{\mathbf{p}} = \mu\dot{\mathbf{v}} = \frac{e_1 e_2 \mathbf{r}}{r^3}$$

($e_1 e_2 > 0$). According to (82.13), the total energy of the bremsstrahlung is

$$\Delta\mathcal{E} = \frac{2\mu^2}{3c^3}\left(\frac{e_1}{m_1} - \frac{e_2}{m_2}\right)^2 \int_{-\infty}^{\infty} \dot{v}^2\,dt = \frac{2}{3c^3}\left(\frac{e_1}{m_1} - \frac{e_2}{m_2}\right)^2 \int_{-\infty}^{\infty} \frac{dt}{r^4}. \tag{1}$$

For a head-on collision the relative velocity v is determined from

$$\frac{\mu v^2}{2} + \frac{e_1 e_2}{r} = \frac{\mu v_\infty^2}{2},$$

where v_∞ is the velocity at infinity. Substituting $dt = dr/v$ in the integral, replacing the integral over t by an integral over r from ∞ to $r_{\min} = 2e_1 e_2/\mu v_\infty^2$ and from $r_{\min}$ to ∞,

$$\int_{-\infty}^{\infty} \frac{dt}{r^4} = 2\int_{r_{\min}}^{\infty} \frac{dr}{r^4\sqrt{(v_\infty^2 - 2e_1 e_2/\mu r)}}.$$

Computing the integral, we find

$$\Delta\mathcal{E} = \frac{8\mu^3 v_\infty^5}{45c^3 e_1 e_2}\left(\frac{e_1}{m_1} - \frac{e_2}{m_2}\right)^2.$$

Problem 3. Determine the total radiation during the passage of one charge past another, if the velocity is so large (though still small compared with c) that the deviation from straight-line motion can be assumed to be small.

Solution. The angle of deflection is small if $\mu v^2 \gg e_1 e_2/\varrho$ (i.e. the kinetic energy $\mu v^2/2$ is large compared with the potential energy, of order $e_1 e_2/\varrho$). For rectilinear motion with speed v, $r = \sqrt{(\varrho^2+v^2t^2)}$, where ϱ is the impact parameter. Substituting in formula (1) of Problem 2 and calculating the integral, we find

$$\Delta\mathscr{E} = \frac{\pi(e_1e_2)^2}{3vc^3\varrho^3}\left(\frac{e_1}{m_2}-\frac{e_2}{m_2}\right)^2.$$

PROBLEM 4. Find the formula for the spectral resolution of bremsstrahlung in the limit of low frequencies.†

SOLUTION. In the integral

$$\ddot{\mathbf{d}}_\omega = \int_{-\infty}^{\infty} \ddot{\mathbf{d}}(t)e^{i\omega t}\,dt = \mu\left(\frac{e_1}{m_1}-\frac{e_2}{m_2}\right)\int_{-\infty}^{\infty}\dot{\mathbf{v}}e^{i\omega t}\,dt,$$

the acceleration $\dot{\mathbf{v}}$ is significantly different from zero only during a time interval $\sim\tau$. Therefore for frequencies $\omega \ll 1/\tau$ we may assume that in the integrand $\omega t \ll 1$ and accordingly put $e^{i\omega t} \approx 1$. Then

$$\ddot{\mathbf{d}}_\omega = \mu\left(\frac{e_1}{m_1}-\frac{e_2}{m_2}\right)\int_{-\infty}^{\infty}\dot{\mathbf{v}}\,dt = \left(\frac{e_1}{m_1}-\frac{e_2}{m_2}\right)\Delta\mathbf{p},$$

where $\Delta\mathbf{p}$ is the change in the momentum of the relative motion $\mathbf{p} = \mu\mathbf{v}$ because of the collision. According to (82.14) the energy radiated in the frequency range $d\omega$ is

$$d\mathscr{E}_\omega = \frac{2}{3\pi c^3}\left(\frac{e_1}{m_1}-\frac{e_2}{m_2}\right)^2(\Delta\mathbf{p})^2\,d\omega.$$

We note that this distribution is independent of frequency, i.e. $d\mathscr{E}_\omega/d\omega$ tends to a constant limit as $\omega \to 0$.

PROBLEM 5. Determine the intensity of the radiation from a charge moving in a circular path in a constant uniform magnetic field.

SOLUTION. From formula (82.9) we find

$$I = \frac{2e^4H^2v^2}{3m^2c^5}.$$

§83. Radiation from a rapidly moving charge

Now let us consider a charged particle moving in an external field with a velocity which is not small compared with the velocity of light. To solve this problem, it is convenient to use the Lienard–Wiechert

† In the spectral resolution of the bremsstrahlung, the main part of the intensity is at frequencies $\omega \sim 1/\tau$, where τ is of the order of the duration of the collision. Correspondingly, by small frequencies we here mean $\omega \ll 1/\tau$.

expressions for the fields, (80.8) and (80.9). At large distances we must retain only the term of lowest order in $1/R$ [the second term in (80.8)]. Introducing the unit vector **n** in the direction of the radiation ($\mathbf{R} = \mathbf{n}R$), we get the formulae

$$\mathbf{E} = \frac{e}{c^2R}\frac{\mathbf{n}\times\{(\mathbf{n}-\mathbf{v}/c)\times\mathbf{w}\}}{(1-\mathbf{n}\cdot\mathbf{v}/c)^3}, \quad \mathbf{H} = \mathbf{n}\times\mathbf{E}, \tag{83.1}$$

where all the quantities on the right-hand sides of the equations refer to the retarded time $t' = t-R/c$.

The intensity radiated into the solid angle do is proportional to E^2. The angular distribution of the radiation is, in general, quite complicated. In the ultra-relativistic case, $1-v/c \ll 1$, it has a characteristic property which is related to the presence of high powers of the difference $1-\mathbf{v}\cdot\mathbf{n}/c$ in the denominators: the intensity is large within the narrow range of angles in which the difference $1-\mathbf{v}\cdot\mathbf{n}/c$ is small. Denoting by θ the small angle between **n** and **v**, we have

$$1-\frac{v}{c}\cos\theta \approx 1-\frac{v}{c}+\frac{\theta^2}{2} \approx \tfrac{1}{2}\left(1-\frac{v^2}{c^2}+\theta^2\right). \tag{83.2}$$

This difference is small if

$$\theta \sim \sqrt{[1-(v^2/c^2)]}. \tag{83.3}$$

Thus an ultra-relativistic particle radiates mainly along the direction of its own motion, within the small range (83.3) of angles around the direction of its velocity.

The amount of energy radiated during time dt into the element of solid angle do is

$$\left(\frac{c}{4\pi}E^2R^2\,\mathrm{d}o\right)\mathrm{d}t. \tag{83.4}$$

In calculating the intensity of the radiation we must now distinguish between two ways of defining it.

In (83.4) dt is the time interval at the field point, so that the expression in parentheses is the intensity defined as the energy of the radiation received by the observer in unit time. But because of the retardation during the propagation of the wave from the radiating particle to the field point, the interval dt is not the same as the time interval dt' during which the energy (83.4) was radiated by the moving particle. According to (80.7) we have

$$dt = \frac{\partial t}{\partial t'}\,dt' = \left(1-\frac{\mathbf{n}\cdot\mathbf{v}}{c}\right)dt'. \tag{83.5}$$

If we define the intensity as the energy radiated by the particle per unit time, it is therefore equal to

$$dI = \frac{c}{4\pi}E^2\left(1-\frac{\mathbf{n}\cdot\mathbf{v}}{c}\right)R^2\,do. \tag{83.6}$$

For $v \ll c$ (as was assumed in §82) the factor $1-\mathbf{n}\cdot\mathbf{v}/c$ can be replaced by unity, so that the two definitions of the intensity coincide.

PROBLEM

Determine the intensity of the radiation from an ultra-relativistic particle, moving in a circular path in a constant uniform magnetic field.

SOLUTION. For the case where the acceleration and the velocity are at right angles, the calculation from (83.1) and (83.6) gives

$$dI = \frac{e^2w^2}{4\pi c^3}\left\{\frac{1}{[1-(v/c)\cos\theta]^3}-\frac{(1-v^2/c^2)\sin^2\theta\cos^2\phi}{[1-(v/c)\cos\theta]^5}\right\}do,$$

where θ is the angle between $\mathbf{n}$ and $\mathbf{v}$, and ϕ is the azimuthal angle of the vector $\mathbf{n}$ with respect to the plane passing through $\mathbf{v}$ and $\mathbf{w}$. In the ultra-relativistic case the small-angle region is the most important. In the region of small θ,

$$dI = \frac{2e^2w^2}{\pi c^3}\left\{\frac{1}{[1-(v^2/c^2)+\theta^2]^3}-\frac{4(1-v^2/c^2)\theta^2}{[1-(v^2/c^2)+\theta^2]^5}\cos^2\phi\right\}do,$$

where the element of solid angle $do = \sin\theta\,d\theta\,d\phi \approx \theta\,d\theta\,d\phi$. Because of the rapid convergence of the integral over θ, in calculating the total intensity we can extend this integral from 0 to ∞. The result is

$$I = \frac{2e^4H^2}{3m^2c^3[1-(v^2/c^2)]}.$$

In this formula we have inserted the expression for the acceleration during circular motion in a magnetic field H:

$$w = \frac{evH}{mc}\sqrt{\left(1-\frac{v^2}{c^2}\right)} \approx \frac{eH}{m}\sqrt{\left(1-\frac{v^2}{c^2}\right)}.$$

§84. Radiation damping

The radiation of electromagnetic waves by moving charges results in a loss of energy by them. The reaction of this loss on the motion of the charges can be described by introducing corresponding "frictional forces" **f** into the equations of motion.

Let us consider a system of charges executing a stationary motion with non-relativistic velocities ($v \ll c$). The average loss of energy from the system (per unit time) is equal to the average intensity of the radiation (82.10). We choose the forces **f** so that this energy loss can be represented as the average work of these forces. The work of the force **f** per unit time is equal to the product $\mathbf{f}\cdot\mathbf{v}$, where **v** is the velocity of the particle. Thus we must have

$$\sum_a \overline{\mathbf{f}_a\cdot\mathbf{v}_a} = -\frac{2}{3c^3}\overline{\ddot{\mathbf{d}}^2} \tag{84.1}$$

(where the sum goes over all the particles of the system).

It is easy to see that this requirement is satisfied by the forces

$$\mathbf{f}_a = \frac{2e_a}{3c^3}\dddot{\mathbf{d}}. \tag{84.2}$$

For we have

$$\sum_a \mathbf{f}_a\cdot\mathbf{v}_a = \frac{2}{3c^3}\dddot{\mathbf{d}}\cdot\sum_a e_a\mathbf{v}_a = \frac{2}{3c^3}\dddot{\mathbf{d}}\cdot\dot{\mathbf{d}} = \frac{2}{3c^3}\frac{\mathrm{d}}{\mathrm{d}t}(\ddot{\mathbf{d}}\cdot\dot{\mathbf{d}}) - \frac{2}{3c^3}\ddot{\mathbf{d}}^2.$$

On averaging, the first term, which contains a total time derivative, vanishes (cf. the footnote on p. 205), and we arrive at (84.1). The forces (84.2) are called *radiation damping* or *Lorentz frictional forces.*

Radiation damping also occurs for the motion of a single particle in an external field. In this case $\dddot{\mathbf{d}} = e\ddot{\mathbf{v}}$, and the equation of motion, including the forces (84.2), has the form

$$m\dot{\mathbf{v}} = e\mathbf{E} + \frac{e}{c}\mathbf{v}\times\mathbf{H} + \frac{2e^2}{3c^3}\ddot{\mathbf{v}}. \tag{84.3}$$

Nevertheless, we must keep in mind that the description of the action of the charge "on itself" with the aid of the damping force is somewhat unsatisfactory, and contains contradictions. For, in the absence of an external field, equation (84.3) reduces to

$$m\dot{\mathbf{v}} = \frac{2e^2}{3c^3}\ddot{\mathbf{v}}.$$

This equation has, in addition to the trivial solution $\mathbf{v}$ = constant, another solution in which the acceleration $\dot{\mathbf{v}}$ is proportional to exp $(3mc^3t/2e^2)$, that is, increases indefinitely with the time. This means, for example, that a charge passing through any field, upon emergence from the field, would have to be infinitely "self-accelerated". The absurdity of this result is evidence for the limited applicability of formula (84.3).

One can raise the question of how electrodynamics, which satisfies the law of conservation of energy, can lead to the absurd result that a free charge increases its energy without limit. Actually the root of this difficulty lies in the earlier remarks (§60) concerning the infinite electromagnetic "intrinsic mass" of elementary particles. When in the equation of motion we write a finite mass for the charge, then in doing this we essentially assign to it formally an infinite negative "intrinsic mass" of non-electromagnetic origin, which together with the electromagnetic mass should result in a finite mass for the particle. Since, however, the subtraction of one infinity from another is not an entirely correct mathematical operation, this leads to a series of further difficulties, among which is the one mentioned here.

Since the damping force thus leads to contradictory results, expression (84.2) is valid only if this force is small compared with the force exerted on the charge by the external field.

§85. Scattering by free charges

If an electromagnetic wave is incident on a system of charges, then under its action the charges are set in motion. This motion in turn produces radiation in all directions; there occurs, we say, a *scattering* of the original wave.

The scattering is conveniently characterised by the ratio of the amount of energy emitted by the scattering system in a given direction per unit time, to the energy flux density of the incident radiation. This ratio has the dimensions of area, and is called the *scattering cross-section* (cf. §15).

Let $\mathrm{d}I$ be the energy radiated by the system into solid angle $\mathrm{d}o$ per unit time for an incident wave with Poynting vector $\mathbf{S}$. Then the cross-section for scattering (into the solid angle $\mathrm{d}o$) is

$$\mathrm{d}\sigma = \overline{\mathrm{d}I}/\overline{S} \tag{85.1}$$

(the bar over a symbol means a time average). The integral σ of $\mathrm{d}\sigma$ over all directions is the *total* scattering cross-section.

Let us consider the scattering produced by a free charge at rest. Suppose there is incident on this charge a plane monochromatic linearly polarised wave. Its electric field can be written in the form

$$\mathbf{E} = \mathbf{E}_0 \cos(\omega t - \mathbf{k}\cdot\mathbf{r} + \alpha).$$

We shall assume that the velocity acquired by the charge under the influence of the incident wave is small compared with the velocity of light (which is almost always the case). Then we can consider the force acting on the charge to be $e\mathbf{E}$, while the force $(e/c)\mathbf{v}\times\mathbf{H}$ due to the magnetic field can be neglected. In this case we can also neglect the effect of the displacement of the charge during its oscillations under the influence of the field. If the charge executes oscillations about the origin, then we can assume that the field which acts on the charge is at all times the same as that at the origin, that is,

$$\mathbf{E} = \mathbf{E}_0 \cos(\omega t + \alpha).$$

Since the equation of motion of the charge is

$$m\ddot{\mathbf{r}} = e\mathbf{E}$$

and its dipole moment $\mathbf{d} = e\mathbf{r}$, then

$$\ddot{\mathbf{d}} = e^2\,\mathbf{E}/m. \tag{85.2}$$

To calculate the scattered radiation, we use formula (82.7) for dipole radiation (this is justified, since the velocity acquired by the charge is assumed to be small). We also note that the frequency of the wave radiated by the charge (i.e. scattered by it) is clearly the same as the frequency of the incident wave.

Substituting (85.2) in (82.7), we find

$$\mathrm{d}I = \frac{e^4}{4\pi m^2 c^3} (\mathbf{E}\times\mathbf{n})^2 \, \mathrm{d}o.$$

On the other hand, the Poynting vector of the incident wave is

$$S = cE^2/4\pi. \tag{85.3}$$

From this we find, for the cross-section for scattering into the solid angle do,

$$\mathrm{d}\sigma = (e^2/mc^2)^2 \sin^2\theta \, \mathrm{d}o, \tag{85.4}$$

where θ is the angle between the direction of scattering (the vector $\mathbf{n}$), and the direction of the electric field $\mathbf{E}$ of the incident wave. We see that the scattering cross-section of a free charge is independent of frequency.

To determine the total cross-section σ, we choose the polar axis along $\mathbf{E}$. Then $\mathrm{d}o = \sin\theta \, \mathrm{d}\theta \, \mathrm{d}\phi$; substituting this and integrating with respect to θ from 0 to π, and over ϕ from 0 to 2π, we find

$$\sigma = \frac{8\pi}{3}\left(\frac{e^2}{mc^2}\right)^2. \tag{85.5}$$

(This is *Thomson's formula*.)

PROBLEMS

PROBLEM 1. Find the cross section dσ for the scattering of an unpolarised wave (natural light).

SOLUTION. We must average (85.4) over all directions of the vector $\mathbf{E}$ in the plane perpendicular to the direction of propagation of the incident wave (the direction of the wave vector $\mathbf{k}$). We choose the z axis along the direction of $\mathbf{k}$, and the x axis along $\mathbf{E}$. Then the cosine of the angle θ between the directions of $\mathbf{n}$ and $\mathbf{E}$, i.e. the projection of the unit vector $\mathbf{n}$ on the x axis, is $\cos\theta = \sin\vartheta\cos\phi$, where ϑ and ϕ are the polar angle and azimuth of the direction $\mathbf{n}$. Averaging over all directions of $\mathbf{E}$ in the plane perpendicular to $\mathbf{k}$ is equivalent to averaging over the azimuth ϕ. We find

$$\overline{\sin^2\theta} = 1 - \tfrac{1}{2}\sin^2\vartheta = \tfrac{1}{2}(1+\cos^2\vartheta),$$

and, substituting in (85.4), we obtain

$$d\sigma = \tfrac{1}{2}(e^2/mc^2)^2(1+\cos^2\vartheta)\, do.$$

PROBLEM 2. Find the frequency ω' of light scattered by a moving charge.

SOLUTION. In the frame of reference where the particle is at rest (the rest frame of the particle), the frequency of the light will not be changed on scattering: $\omega'=\omega$. This relation can be written in invariant form:

$$k'_\mu u^\mu = k_\mu u^\mu,$$

where k^μ and k'^μ are the wave four-vectors of the incident and scattered light, and u^μ is the four-velocity of the particle (the only non-zero component in the rest frame is $u^0 = 1$). Writing out this equality in an arbitrary frame of reference (in which the particle moves with velocity $\mathbf{v}$), we find

$$\omega'\left(1-\frac{v}{c}\cos\theta'\right) = \omega\left(1-\frac{v}{c}\cos\theta\right),$$

where θ and θ' are the angles made by the directions of the incident and scattered waves with the direction of $\mathbf{v}$.

PROBLEM 3. Find the cross-section for scattering of a linearly polarised wave by a space oscillator—a charge executing small oscillations (under the influence of some elastic force) with frequency ω_0. Include the effects of radiation damping.

SOLUTION. We write the equation of motion of the oscillator in the incident wave in the form

$$\ddot{\mathbf{r}}+\omega_0^2\mathbf{r} = \frac{e}{m}\mathbf{E}_0 e^{-i\omega t}+\frac{2e^2}{3mc^3}\dddot{\mathbf{r}}.$$

In the damping force (the second term on the right) we can substitute approximately $\dddot{\mathbf{r}} = -\omega_0^2\dot{\mathbf{r}}$; then we find

$$\ddot{\mathbf{r}}+\gamma\dot{\mathbf{r}}+\omega_0^2\mathbf{r} = (e/m)\,\mathbf{E}_0\, e^{-i\omega t}, \qquad \gamma = (2e^2/3mc^3)\omega_0^2.$$

From this we obtain for forced oscillations

$$\mathbf{r} = \frac{e}{m}\mathbf{E}_0\frac{e^{-i\omega t}}{\omega_0^2-\omega^2-i\omega\gamma}.$$

The further calculation is done as in the text (the computation of the average of the square of a complex quantity should be done as in the footnote on p. 219). The result for the scattering cross-section is

$$\sigma = \frac{8\pi}{3}\left(\frac{e^2}{mc^2}\right)^2\frac{\omega^4}{(\omega_0^2-\omega^2)^2+\omega^2\gamma^2}.$$

§86. Scattering by a system of charges

The scattering of electromagnetic waves by a system of charges differs from the scattering by a single charge (at rest), first of all in the fact that, because of the presence of internal motion of the charges in the system, the frequency of the scattered radiation can be different from the frequency of the incident wave. Namely, in the spectral resolution of the scattered wave there appear, in addition to the frequency ω of the incident wave, frequencies ω' differing from ω by one of the eigenfrequencies of motion of the scattering system. The scattering with changed frequency is called *incoherent* (or *Raman*) scattering, in contrast to the *coherent* scattering without change in frequency.

Assuming that the field of the incident wave is weak, we can represent the current density in the form $\mathbf{j} = \mathbf{j}_0 + \mathbf{j}'$, where $\mathbf{j}_0$ is the current density in the absence of the external field, and $\mathbf{j}'$ is the change in the current under the action of the incident wave. Correspondingly, the vector potential (and other quantities) of the field of the system also has the form $\mathbf{A} = \mathbf{A}_0 + \mathbf{A}'$, where $\mathbf{A}_0$ and $\mathbf{A}'$ are determined by the currents $\mathbf{j}_0$ and $\mathbf{j}'$. $\mathbf{A}'$ describes the wave scattered by the system, and is given in terms of $\mathbf{j}'$ by formula (81.2):

$$\mathbf{A}' = \frac{1}{cR_0} \int \mathbf{j}'_{t-R_0/c+\mathbf{r}\cdot\mathbf{n}/c}\, \mathrm{d}V. \qquad (86.1)$$

Let us consider the two limiting cases where the frequency ω of the scattered waves is small or large compared with the fundamental modes of the system. The latter are of order $\omega_0 \sim v/a$, where v is the velocity of the charges in the system and a its size. We shall also assume that the velocities $v \ll c$.

We start with the case where

$$\omega \ll \omega_0 \sim v/a. \qquad (86.2)$$

The scattering can contain both coherent and incoherent parts, but we shall here consider only the coherent scattering.

Under the condition (86.2) we can make the same approximations in formula (86.1) as we did in §82. In other words, the scattered radiation will be dipole radiation. If this is so, then the intensity of its spectral component with frequency ω will be proportional to the square of the Fourier component $|\ddot{\mathbf{d}}'_\omega|^2 = \omega^4 |\mathbf{d}'_\omega|^2$, where $\mathbf{d}'$ is the change in the dipole moment under the influence of the incident wave.

If the total charge of the system is zero (a neutral atom or molecule), then for $\omega \to 0$, $\mathbf{d}'_\omega$ approaches a constant limit (if the sum of the charges were different from zero, then for $\omega = 0$, i.e. for a constant field, the system would begin to move as a whole). Therefore for low frequencies ω we can consider $\mathbf{d}'_\omega$ as independent of frequency. Then the intensity of the scattered wave, and, with it, the cross-section for scattering are proportional to the fourth power of the frequency of the incident radiation:

$$\sigma_{\text{coh}} = \text{constant} \times \omega^4. \tag{86.3}$$

We proceed now to the opposite case of high frequencies:

$$\omega \gg \omega_0 \sim v/a. \tag{86.4}$$

From this condition, the periods of the motion of the charges in the system are large compared with the period of the wave. Therefore during a time interval of the order of the period of the wave, the motion of the charges in the system can be considered uniform. This means that, in considering the scattering of short waves, we need not take into account the interaction of the charges in the system with each other, that is, we can consider them as free.

Thus in calculating the velocity $\mathbf{v}'$ acquired by a charge in the field of the incident wave, we can consider each of the charges in the system separately, and write for it an equation of motion of the form

$$m \, d\mathbf{v}'/dt = e\mathbf{E} = e\mathbf{E}_0 \, e^{-i(\omega t - \mathbf{k}\cdot\mathbf{r})},$$

where $\mathbf{k} = (\omega/c)\mathbf{n}$ is the wave vector of the incident wave. The position of the charge is, of course, a function of the time. In the exponent on the right-hand side of this equation the time rate of change of the first term is large compared with that of the second (the first is ω, while the second is of order $kv \sim v(\omega/c) \ll \omega$). Therefore, in integrating the equations of motion, we can consider the $\mathbf{r}$ on the right-hand side as constant. Then

$$\mathbf{v}' = -\frac{e}{i\omega m} \mathbf{E}_0 \, e^{-i(\omega t - \mathbf{k}\cdot\mathbf{r})}. \tag{86.5}$$

For the vector potential of the scattered wave (at large distances from the system), we have, changing (86.1) to a sum,

$$\mathbf{A}' = \frac{1}{cR_0}\sum (e\mathbf{v}')_{t-R_0/c+\mathbf{r}\cdot\mathbf{n}'/c},$$

where $\mathbf{n}'$ is a unit vector in the direction of scattering. Substituting (86.5), we find

$$\mathbf{A}' = -\frac{1}{icR_0\omega}\, e^{-i\omega(t-R_0/c)}\mathbf{E}_0 \sum \frac{e^2}{m}\, e^{-i\mathbf{q}\cdot\mathbf{r}}, \tag{86.6}$$

where $\mathbf{q} = \mathbf{k}'-\mathbf{k}$ is the difference between the wave vector $\mathbf{k} = (\omega/c)\mathbf{n}$ of the incident wave, and the wave vector $\mathbf{k}' = (\omega/c)\mathbf{n}'$ of the scattered wave.† The value of the sum in (86.6) must be taken at the time $t' = t-R_0/c$ (for brevity as usual, we omit the index t' on $\mathbf{r}$); the change of $\mathbf{r}$ in the time $\mathbf{r}\cdot\mathbf{n}'/c$ can be neglected in view of our assumption that the velocities of the particles are small. The absolute value of the vector $\mathbf{q}$ is

$$q = 2(\omega/c)\sin\tfrac{1}{2}\vartheta, \tag{86.7}$$

where ϑ is the scattering angle.

For scattering by an atom (or molecule), we can neglect the terms in the sum in (86.6) which come from the nuclei, because their masses are large compared with the electron mass. Below we shall be looking at just this case, so that we take the factor e^2/m outside the summation sign, and understand by e and m the charge and mass of the electron.

For the field $\mathbf{H}'$ of the scattered wave we find from (81.3):

$$\mathbf{H}' = \frac{\mathbf{E}_0\times\mathbf{n}'}{c^2R_0}\, e^{-i\omega(t-R_0/c)}\frac{e^2}{m}\sum e^{-i\mathbf{q}\cdot\mathbf{r}}. \tag{86.8}$$

The energy flux into an element of solid angle in the direction $\mathbf{n}'$ is

$$\frac{c\,|\mathbf{H}'|^2}{8\pi}R_0^2\,do = \frac{e^4}{8\pi c^3m^2}(\mathbf{E}_0\times\mathbf{n}')^2\left|\sum e^{-i\mathbf{q}\cdot\mathbf{r}}\right|^2 do.$$

† Strictly speaking, the wave vector $\mathbf{k}' = \omega'\mathbf{n}'/c$, where the frequency ω' of the scattered wave may differ from ω. However, in the present case of high frequencies the difference $\omega'-\omega \sim \omega_0$ can be neglected.

Dividing this by the mean energy flux $(c/8\pi)\,|\mathbf{E}_0|^2$ of the incident wave (cf. the footnote on p. 219) and introducing the angle θ between the direction of the field $\mathbf{E}$ of the incident wave and the direction of scattering, we finally obtain the scattering cross-section in the form

$$d\sigma = \left(\frac{e^2}{mc^2}\right)^2 \overline{|\sum e^{-i\mathbf{q}\cdot\mathbf{r}}|^2} \sin^2\theta \, do. \tag{86.9}$$

The bar means a time average, i.e. an average over the motion of the charges in the system; it appears because the scattering is observed over a time interval large compared with the periods of motion of the charges in the system.

For the wavelength of the incident radiation, there follows from the condition (86.4) the inequality $\lambda \ll (c/v)\cdot a$. As for the relative values of λ and a, both the limiting cases $\lambda \gg a$ and $\lambda \ll a$ are possible. In both these cases the general formula (86.9) simplifies considerably.

In the case of $\lambda \gg a$, in the expression (86.9) $\mathbf{q}\cdot\mathbf{r} \ll 1$, since $q \sim 1/\lambda$, and r is of the order of a. Replacing $e^{-i\mathbf{q}\cdot\mathbf{r}}$ by unity in accordance with this, we have

$$d\sigma = \left(\frac{Ze^2}{mc^2}\right)^2 \sin^2\theta \, do; \tag{86.10}$$

that is, the scattering is proportional to the square of the atomic number Z.

We now go over to the case of $\lambda \ll a$. In the square of the sum which appears in (86.9), in addition to the square modulus of each term, which is unity, there appear products of the form $e^{i\mathbf{q}\cdot(\mathbf{r}_1-\mathbf{r}_2)}$. In averaging over the motion of the charges, i.e. over their mutual separations, $\mathbf{r}_1-\mathbf{r}_2$ takes on values in an interval of order a. Since $q \sim 1/\lambda$, $\lambda \ll a$, the exponential factor $e^{i\mathbf{q}\cdot(\mathbf{r}_1-\mathbf{r}_2)}$ is a rapidly oscillating function in this interval, and its average value vanishes. Thus, for $\lambda \ll a$, the scattering cross-section is

$$d\sigma = Z\left(\frac{e^2}{mc^2}\right)^2 \sin^2\theta \, do, \tag{86.11}$$

that is, the scattering is proportional to the first power of the atomic number.

The cross-sections (86.9)–(86.11) contain both coherent and incoherent parts. To determine the coherent scattering cross-section, we must separate the part of the field of the scattered wave that has frequency ω. The expression (86.8) for the field depends on the time through the factor $e^{-i\omega t}$ and also through the time dependence of the sum $\Sigma e^{-i\mathbf{q}\cdot\mathbf{r}}$. This latter dependence also has the consequence that the field of the scattered wave contains not only the frequency ω but other (slightly different) frequencies. The part of the field that has frequency ω (i.e. depends on the time only through the factor $e^{-i\omega t}$) is obtained by taking the time average of $\Sigma e^{-i\mathbf{q}\cdot\mathbf{r}}$. The corresponding expression for the coherent scattering cross-section $d\sigma_{coh}$ differs from the total cross-section $d\sigma$; instead of the average value of the squared modulus of the sum it contains the squared modulus of the average value of the sum:

$$d\sigma_{coh} = \left(\frac{e^2}{mc^2}\right)^2 |F(\mathbf{q})|^2 \sin^2\theta \, do, \tag{86.12}$$

where

$$F(\mathbf{q}) = \Sigma \overline{e^{-i\mathbf{q}\cdot\mathbf{r}}}. \tag{86.13}$$

The function $F(\mathbf{q})$ is called the *atomic form factor*. It is worth noting that it is just the spatial Fourier component of the average charge distribution in the atom, $\bar{\varrho}(\mathbf{r})$:

$$eF(\mathbf{q}) = \int \bar{\varrho}(\mathbf{r})\, e^{-i\mathbf{q}\cdot\mathbf{r}}\, dV. \tag{86.14}$$

This is easily understood if from the start we write the unaveraged density $\varrho(\mathbf{r})$ as a sum of delta functions (cf. (54.1)).

If $\lambda \gg a$, we can again replace $e^{-i\mathbf{q}\cdot\mathbf{r}}$ by unity, so that

$$d\sigma_{coh} = \left(Z\frac{e^2}{mc^2}\right)^2 \sin^2\theta \, do. \tag{86.15}$$

Comparing this with the total cross-section (86.10), we see that $d\sigma_{coh} = d\sigma$, that is, all the scattering is coherent.

If $\lambda \ll a$, then when we average in (86.13) all the terms of the sum (being mean values of rapidly oscillating functions of the time) vanish, so that $d\sigma_{coh} = 0$. Thus in this case the scattering is completely incoherent.

INDEX